OXFORD TEX. _ _

Genetics and the Literary Imagination

GENERAL EDITORS

Elaine Treharne   Greg Walker

Genetics and the Literary Imagination

CLARE HANSON

OXFORD
UNIVERSITY PRESS

OXFORD
UNIVERSITY PRESS

Great Clarendon Street, Oxford, OX2 6DP,
United Kingdom

Oxford University Press is a department of the University of Oxford.
It furthers the University's objective of excellence in research, scholarship,
and education by publishing worldwide. Oxford is a registered trade mark of
Oxford University Press in the UK and in certain other countries

First Edition published in 2020

Impression: 1

Published in the United States of America by Oxford University Press
198 Madison Avenue, New York, NY 10016, United States of America

British Library Cataloguing in Publication Data
Data available

Library of Congress Control Number: 2020930613

ISBN 978-0-19-881328-6 (Hbk.)
ISBN 978-0-19-881334-7 (Pbk.)

Printed and bound in Great Britain by
Clays Ltd, Elcograf S.p.A.

ACKNOWLEDGEMENTS

I am grateful to the Arts and Humanities Research Council for a Science in Culture Exploratory Award, 'Beyond the Gene: Epigenetic Science in Twenty-First Culture', which enabled much of the preliminary thinking for this project. Warm thanks also to the Faculty of Humanities at the University of Southampton for study leave which allowed me to complete a significant portion of this book. Early versions of individual chapters have been presented at the University of Bristol, the University of Exeter, Hamburg University, the University of Leeds, and the University of Warwick and have benefitted from the audiences' generous and perceptive feedback. I have also learnt a great deal from conversations with Ruth Müller, with whom I co-organized a workshop on 'Plasticity and its Limits' at Munich Technical University, while from the biomedical point of view, Mark Hanson has been an indispensable interlocutor throughout.

This book is for my extended family, with love.

Permissions

An earlier version of some of the material in Chapter 5 was first published as 'Epigenetics, Plasticity and Identity in Jackie Kay's *Red Dust Road*', *Textual Practice* 29, 3 (2015), 433–52 and is reprinted by permission of Informa UK.

The Agreed Upon 116 Words from *Kazuo Ishiguro Papers* by Kazuo Ishiguro. Published by the Harry Ransom Center. Copyright c Kazuo Ishiguro. Reproduced by permission of the author c/o Rogers, Coleridge and White Ltd, 20 Powis Mews, London W11 1JN.

The Agreed Upon 186 Words from *Ian McEwan's Archive at the Harry Ransom Center* by Ian McEwan. Published by the Harry Ransom Center. Copyright c Ian McEwan. Reproduced by permission of the author c/o Rogers, Coleridge and White Ltd, 20 Powis Mews, London W11 1JN.

CONTENTS

Introduction: The Secret of Life

On the morning of 28 February 1953, James Watson and Francis Crick finally realized how the base pairs of DNA could fit together in a double helical structure, and at lunchtime on the same day, Crick rushed into the Eagle pub to tell everyone that they had discovered the secret of life.[1] Or rather, that is the story James Watson tells in *The Double Helix*, but Crick always denied that he had said such a thing, and at a meeting at Cold Spring Harbour to mark the centenary of Crick's birth, Watson admitted that he had made the story up for dramatic effect.[2] Both Watson and Crick had a gift for vivid exposition and each played a part in shaping the discursive framework in which DNA was positioned as the logos of life in the years that followed, when the genetic 'code' was cracked and recombinant DNA techniques were developed. Scriptural metaphors linking the language of life with biblical myths of origin became a staple of genetic discourse and the idea that life itself could be read from the genomic text was built into the mandate for the Human Genome Project (HGP) which was established in 1990 under Watson's direction. However, as the choice of the term genome suggests, by then the object of study was shifting from the gene

---

[1] James Watson, *The Double Helix* (London: Penguin Books, 1999), p. 155.
[2] See Matthew Cobb, 'Happy 100th Birthday, Francis Crick (1916–2004)', Why Evolution is True website, available at https://whyevolutionistrue.wordpress.com/2016/06/08/happy-100th-birthday-francis-crick-1916-2004/ [accessed 17 July 2019].

Genetics and the Literary Imagination. Clare Hanson, Oxford University Press (2020). © Clare Hanson.
DOI: 10.1093/oso/9780198813286.001.0001

(as protein-coding DNA) to the genome (as a system of interacting macromolecules). This process has intensified as research in epigenetics has expanded our understanding of the extent to which gene expression is modified by other chemical components of the cell. The genome is now seen as a reactive system whereby the cell responds to its internal and external environments, and the neo-Darwinian 'thought style' associated with Watson and Crick has given way to a postgenomic era in which, as Richardson and Stevens suggest, there is considerable uncertainty about 'the proliferating objects, relationships, and levels' involved in the relationship between DNA and the organism.[3]

This book explores the impact of genetics, genomics, and postgenomics on British fiction over the last four decades, focusing on the challenge posed to novelists by gene-centric neo-Darwinism and examining the recent rapprochement between postgenomic and literary perspectives on human nature. The term neo-Darwinism requires some initial explanation, as its meaning has shifted over time. It was first coined by George Romanes in 1895 to refer to the evolutionary theories of Alfred Russel Wallace and August Weissman and was a slightly derogatory term intended to emphasize the difference between their theories and those of Darwin. Darwin argued that evolution was primarily driven by natural selection operating on particles of inheritance but also allowed for the possibility that such particles, which he called 'gemmules', could be modified by the environment, leading to alternative forms of inheritance. In contrast with Darwin's pluralistic approach, Wallace promoted a rigid selectionism which ruled out the idea, associated with Jean-Baptiste de Lamarck, that characteristics acquired during the life of an organism could be inherited. Weissman took this line further by arguing for a separation between the germ cells which were thought to carry inheritance and the body's somatic cells, a separation which became known as Weissman's barrier and which implied that the germ cells were immutable and impervious to external influences. Weissman also proposed that the sole driver of evolution

---

[3] *Postgenomics: Perspectives on Biology after the Genome*, ed. by Sarah S. Richardson and Hallam Stevens (Durham and London; Duke University Press, 2015), p. 6. Further page references will be given within the main text. Ludwik Fleck's term 'thought style' is used here in preference to Kuhn's 'paradigm' as it captures the sociocultural dimensions of scientific theories.

was spontaneous generation or what would now be called random mutation. Together, Wallace and Weismann had established the core principles of neo-Darwinism, which were that inheritance was particulate, that the units of inheritance were 'sealed' from the environment and that the single force behind evolution was random mutation.[4] These commitments were carried through into the classical genetics of the early twentieth century and also underpinned the 'Modern Synthesis' of the 1930s and 1940s which united evolutionary theory, Mendelian genetics, and population genetics. However, it was in the post-war period that neo-Darwinism mutated into what Eva Jablonka and Marion Lamb term 'molecular neo-Darwinism' (30–4). In this iteration, the gene of the Modern Synthesis became the DNA sequence which codes for protein, in a one-way process whereby in Crick's words, 'DNA makes RNA, RNA makes proteins, and proteins make us'.[5] DNA became the 'master molecule', DNA sequences were characterized as instructions or information, and Francois Jacob and Jacques Monod introduced the concept of a genetic 'programme' which governed cellular processes.[6] As these metaphors suggest, this was a top-down model of inheritance in which traits emerged according to a pre-existing blueprint. While the molecularization of genetics proceeded apace, evolutionary biologists were concerned with a related problem, the level on which natural selection acts. They were particularly exercised by the concept of group selection, which was associated with the rather vague idea that individuals acted for the good of the group or species. The 'good for the species' explanation of altruistic behaviour did not satisfy the evolutionary biologist W.D. Hamilton and in its place he offered the theory of kin selection, whereby organisms can act in a way that damages their own interests if the action nonetheless ensures

---

[4] For an authoritative overview of Darwinism and neo-Darwinism see Eva Jablonka and Marion J. Lamb, *Evolution in Four Dimensions: Genetic, Epigenetic, Behavioural, and Symbolic Variation in the History of Life* (Cambridge, MA and London: MIT Press, 2005), pp. 10–21. Further page references will be given within the main text.

[5] Francis Crick, 'On Protein Synthesis' (1957), quoted in Evelyn Fox Keller, *The Century of the Gene* (Cambridge MA and London: Harvard University Press, 2000), p. 54.

[6] Although Jacob and Monod are usually seen as originating the idea of the genetic programme, Ernst Mayr developed the concept independently. For a discussion of the genesis of the genetic programme see Alexandre E. Peluffo, 'The "Genetic Program": Behind the Genesis of an Influential Metaphor', *Genetics* 200 (2015), 685–96. doi:10.1534/genetics.115.178418

the survival of kin with the same genes. Richard Dawkins took up Hamilton's idea and extended it, arguing that 'the gene's eye view' can help us to understand the evolution of all adaptive traits, and as Jablonka and Lamb suggest, by drawing on Hamilton's evolutionary theories he was able to 'generalize 'the molecular neo-Darwinian approach in his hugely influential *The Selfish Gene* (36).

Things have changed significantly since the completion of the Human Genome Project, which brought the news that human beings had far fewer genes than was anticipated, between 20–25,000 as opposed to a projected 80,000. This invited the question of what the extragenic DNA was for and the ENCODE project which was set up to explore this has found that much of this so-called 'junk' DNA is involved in processes of cellular regulation.[7] Meanwhile, research in epigenetics has rehabilitated the Lamarckian concept of the inheritance of acquired characters, which was considered more-or-less taboo throughout the twentieth century. It is now clear that genomes respond to signals from the environment and that they modify cellular function accordingly. In addition, there has been a return of interest in the role of cooperation and symbiosis in driving evolution, as research on the microbiome has demonstrated the importance of mutual dependencies between organisms. Again, this is in sharp contrast with the neo-Darwinian model of competitive struggle between genes and organisms. Finally, there has been a decisive move to a conceptualization of the genome as a dynamic system which is nested within other dynamic systems, and a renewed interest in the autopoietic theories developed by Humberto Maturana and Francisco Varela, who see living systems as self-organizing and structurally coupled with their environment.[8] To an extent, then, postgenomics can be seen as the return of the repressed of neo-Darwinism, although as Richardson and Stevens warn, it does not represent a clean break with older modes of research and there are many continuities between twentieth-century genetics and the postgenomic present (3–4). Notably, the reductionism and determinism of neo-Darwinism continues to structure much research, especially that

---

[7] See the ENCODE Project Consortium, 'An Integrated Encyclopedia of DNA Elements in the Human Genome', *Nature* 489, 7414 (2012): 57–74.

[8] See Humberto R. Maturana and Francisco J. Varela, *Autopoiesis and Cognition: The Realization of the Living* (Dortrecht: D. Reidel, 1980).

which has a biomedical orientation: in the context of the need for therapeutic applications, there is pressure to identify specific causal mechanisms at the molecular level. Yet although neo-Darwinian assumptions continue to inflect postgenomic practice, the holistic theories associated with postgenomics have a transformative potential which has been recognized across disciplines, including philosophy and literary studies.

This brings us to the assumptions that underlie this book and its understanding of the relationship between literature and science, which has traditionally been conceptualized in two ways. The first is the 'influence' approach, effectively a one-way model which traces the impact of science on literature. This tends to assume a hierarchical relationship between science and literature, granting epistemological and ontological priority to science, and as Rachel Crossland notes, while its exponents are happy to argue for the influence of science on literature, they are reluctant to make any stronger causal claim than that literature may 'anticipate' scientific developments.[9] The other approach explores parallels between literature and science, which it sees as rooted in a common cultural matrix or shared milieu. In this model, it is assumed that writers and scientists share what Gillian Beer has called 'the moment's discourse' and that in consequence literature and culture do not simply reflect scientific findings but are engaged in what Sally Shuttleworth terms 'a dynamic, reciprocal set of relations with scientific practice and the development of scientific ideas'.[10] Like most books concerned with literature and science, the present study combines the two approaches and suggests that in practice they are difficult to disentangle. In relation to neo-Darwinism, for example, this thought style clearly has a long history in evolutionary biology and was consolidated in the heroic age of molecular biology: in this sense, the impetus comes from science and literature responds. However, as Beer argues, literary responses to science are by no means passive and are best understood in terms of transformation rather than translation

---

[9] Rachel Crossland, *Modernist Physics: Waves, Particle and Relativities in the Writings of Virginia Woolf and D.H. Lawrence* (Oxford English Monographs) (Oxford: Oxford University Press, 2018, p. 5.

[10] Gillian Beer, *Open Fields: Science in Cultural Encounter* (Oxford: Oxford University Press, 1996), p. 171; Sally Shuttleworth, 'Life in the Zooniverse: Working with Citizen Science', *Journal of Literature and Science* 10 (2017), 46–51 (p. 46).

from one sphere to another. An example in relation to neo-Darwinism would be Ian McEwan's mobilization of concepts from evolutionary psychology to radically reconfigure our understanding of the literary trope of the unreliable narrator. At a macro-level, it is also clear that scientific theories are forged in the context of broader cultural narratives and respond, to a greater or lesser degree, to their social and historical moment. E.O. Wilson's sociobiological theories, for example, with their bleak view of human nature, are imbued with a sense of angst which can be traced back to his reading of existentialist literature and to a wider post-war climate which demanded reflection on the banality of evil, to invoke Arendt's resonant phrase.

The relationship between literature and biology is especially close, as of all the sciences, biology touches most directly on our self-understanding and the way in which we envisage social relations. Indeed, evolutionary thought was from the outset closely entwined with social theory: Darwin was initially inspired by reading the political economy of Thomas Malthus and Karl Marx was in turn influenced by Darwin's theories, seeing a connection between natural history and class struggle. After the publication of *The Origin of Species*, the social implications of evolutionary theory were widely debated by figures such as Thomas Henry Huxley, who argued that the qualities that fit us for success in the 'struggle for existence' are in conflict with law, morality, and ethics but that we must be prepared to combat such evolutionary pressures.[11] On the other side of this argument, Herbert Spencer, a Lamarckian whose work is often linked with Social Darwinism, contended that competitive struggle was necessary in order to maximize individual self-development. In the twentieth century, particularly after the revelation of Nazi atrocities, genetics sought to position itself as a neutral science working for the benefit of all mankind but with the advent of neo-Darwinism, the controversies that had surrounded the publication of *The Origin of Species* were, in Gillian Beer's words, 'raised anew in a more immediate form'.[12] The issues were similar in

---

[11] See T.H. Huxley, *Evolution and Ethics*, Prolegomena, 1894, Project Gutenberg, https://www.gutenberg.org/files/2940/2940-h/2940-h.htm [accessed 17 July 2019].
[12] Gillian Beer, *Darwin's Plots: Evolutionary Narrative in Darwin, George Eliot and Nineteenth-Century Fiction*, 2nd edn (Cambridge: Cambridge University Press, 2000), p. xxiii.

that they turned on the conflict between evolutionary processes and the ethical ideals of humanity and in this sense, Dawkins reiterates T.H. Huxley's point when he argues in the peroration to *The Selfish Gene* that human beings should discuss ways of 'deliberately cultivating and nurturing pure, disinterested altruism' in order that we can 'rebel against the tyranny of the selfish genes'.[13] The issues were more immediate because neo-Darwinism offered a perspective on evolution that differed from that of Darwin in being both reductionist and deterministic (as many have pointed out, Darwin was no neo-Darwinist). This led to a sharper sense of the conflict between ethics and evolutionary biology. The logic of neo-Darwinism is that human behaviour is impelled by genetic self-interest and to the extent that this is so, human agency is circumscribed. Both Wilson and Dawkins urge a kind of ethical resistance to genetic imperatives, but this is a possibility for which they can find little evolutionary justification. The implications for personal and social relations and hence for the literary categories of character and plot are a recurring theme in the literary texts which engage with their work.

The relationship between literature and biology, then, is one that has involved both reciprocity and conflict and it is also, crucially, a mediated relationship. Most writers do not access scientific knowledge by reading original papers, which are subject to strict protocols surrounding the representation and discussion of results. Instead, they access the sciences through the medium of popular science, a hybrid genre which mediates between science and the wider public. The origin of this form is usually dated to the scientific revolution of the eighteenth century and historians of science have argued that it expanded in the second half of the nineteenth century due to the increasing specialization of science, which led to a perceived need for scientific findings to be 'translated' for the general public. As science has become ever more specialized and fragmented, popular science has stepped in to bridge the gaps between scientific sub-disciplines as well as between science and the general public. As historians of science also contend, popular science need not entail dumbing down but can open new intellectual territory, as for example connections are made between different areas

[13] Richard Dawkins, *The Selfish Gene*, 2nd edn (Oxford: Oxford University Press, 1989), pp. 200–1. Further page references will be given within the main text.

of specialism. And as Richard Dawkins points out, the task of making science accessible can prompt the author to 'push novelty of language and metaphor' to the point where 'a new way of seeing' emerges which can 'in its own right make an original contribution to science' (xi). This is particularly true of the books which appeared during the popular science boom of the period between the mid-seventies and the new millennium. Dawkins' *The Selfish Gene* and *The Extended Phenotype*, Stephen Hawkins' *A Brief History of Time* and Steven Pinker's *The Language Instinct* not only synthesized existing knowledge but offered a distinctive interpretation of the writer's own field. In this sense, there is no clear-cut separation between science and its popularization but as Salman Rushdie notes in a review of *A Brief History of Time*, there has nonetheless been an increasing tendency to make grandiose claims in popular science: 'these days' he writes 'the creation of Creation is primarily the work of scientific, rather than literary or theological, imaginations'.[14] Popular science in the first half of the twentieth century, as Peter Bowler has shown, was sober in style and largely aimed at self-improvement but there was a shift in genre expectations as a new kind of popular science book emerged in the 1960s and 1970s, sometimes with a TV tie-in, David Attenborough's *Life on Earth* being the classic example.[15] In this context, it became more common for popular science to make claims about the importance of science for human self-understanding as well as plumbing the secrets of the universe.

This shift to generalizing claims encouraged the perception that biology was moving into the territory of the humanities and undermining its foundations. Specifically, popular books by Dawkins, Wilson, Ridley, and others explained human values and human behaviour in terms of algorithmic processes, in a move which radically decentred the humanist subject. When Dawkins represented human beings as 'lumbering robots' created for the benefit of self-replicating genes, he was striking a blow at human self-understanding which in some respects resembled the anti-humanism of poststructuralism and

---

[14] Salman Rushdie, *Imaginary Homelands: Essays and Criticism 1981–1991* (London: Penguin, 1991), p. 262.
[15] See Peter J. Bowler, *Science for All: The Popularisation of Science in Early Twentieth-Century Britain* (Chicago: Chicago University Press, 2009), p. 268.

postmodernism. As Rosi Braidotti has argued, postmodernism offered a powerful critique of the unified subject construed in terms of autonomy and self-determination, while Derridean poststructuralism construed the subject as an effect of *différance* and as a crossing point of discourses.[16] Neo-Darwinism similarly challenges the idea of human self-determination and frames the human as a crossing point of *genetic* inscriptions. However, here the parallels with poststructuralism and postmodernism end, as in place of a world of shifting signifiers and mobile identities, neo-Darwinism offers a decisive picture of human nature and history, a Lyotardian 'grand narrative' backed by the epistemological authority of science. The category of the human may have been decentred but in place of our traditional self-conception, neo-Darwinism tells us that much of our behaviour stems from dispositions which evolved in the ancestral past. In this view, the origin and end of life is reproduction and what we might have thought of as ethical behaviour, for example parental love or kindness to a stranger, is reconfigured in terms of genetic self-interest (now termed kin altruism and reciprocal altruism). As Steven Pinker acknowledges, this borders on a tragic vision and it is one which displaces long-held beliefs about 'the perfectibility of man'.[17] Nonetheless, there was a public appetite for this perspective, which according to A.S. Byatt was perceived as liberating precisely because it challenged the orthodoxies of poststructuralism and postmodernism.[18]

Novelists responded to this incursion into the territory of the humanities in multiple and strategic ways. Although they were not able to query the epistemological or ontological status of the biological arguments, they could point to the inflated and potentially misleading rhetoric employed by writers like Dawkins and Wilson. As we shall see, the epideictic, declamatory tone of neo-Darwinian popular science, evident especially in the work of popularizers such as Matt Ridley, is the object of fictional pastiche which draws attention to the clichés that

---

[16] Rosi Braidotti, *The Posthuman* (Cambridge: Polity Press, 2013), p. 37. Further page references will be given within the main text.
[17] Steven Pinker, *The Blank Slate: The Modern Denial of Human Nature* (London: Penguin, 2003), p. 287.
[18] A.S Byatt, 'Faith in Science', *Prospect Magazine* 20 November 2000, https://www.prospectmagazine.co.uk/magazine/faithinscience [accessed 18 July 2019].

can inform such writing.[19] Writers were also well-placed to draw attention to the problematic use of literary techniques, particularly metaphor, in popular science. Metaphors are intrinsic to scientific thought, enabling novel conceptualizations of the material world, but become problematic when they are used in the crossover genre of popular science. The fiction discussed here explores the multiple and often contradictory meanings of the metaphors most often associated with genetics, such as information, code, and programme. However, the novel's most significant contribution has been to draw attention to the limitations of neo-Darwinism's third-person perspective on human nature. The Pulitzer prize-winning novelist Marilynne Robinson has written eloquently about this, arguing that neo-Darwinians offer an impoverished view of humanity because they exclude 'the felt life of the mind' and the 'experience and testimony of humankind' from their evidential base.[20] To compensate for this, they extrapolate from limited scientific data to make what Robinson calls 'parascientific' claims about humanity and its purposes. For Robinson, parascience (as exemplified by the work of E.O. Wilson in particular) is a genre that proceeds from 'the science of its moment, from a genesis of human nature in primordial life to a set of general conclusions about what our nature is and must be, together with the ethical, political, economic and/or philosophical implications to be drawn from these conclusions' (32–3). Subjectivity is invoked only to be dismissed as a source of illusions which disguise the true purpose of our behaviour from us: so for Wilson, love, courage, and generosity are 'illusory sensations', merely the means by which the genes that have colonized us manipulate us for their purposes (61). For Robinson, in contrast, subjectivity is 'the ancient haunt of [. . .] long, long thoughts' and the most important resource for understanding what it means to be human. In this context,

---

[19] The uneasy combination of hard science and clichéd rhetoric that marks much popular science writing is satirized in Zadie Smith's *White Teeth*, where a geneticist co-authors a popular book with a novelist, creating a 'split-level, high/low culture book' (Zadie Smith, *White Teeth* (London: Penguin, 2001), p. 416). For an excellent analysis of Smith's treatment of the fictional dimension of genetics see Josie Gill, 'Science and Fiction in Zadie Smith's *White Teeth*', *Journal of Literature and Science* 6 (2013), 17–28.

[20] Marilynne Robinson, *Absence of Mind: The Dispelling of Inwardness from the Modern Myth of the Self* (New Haven and London: Yale University Press, 2010), p. 35. Further page references will be given within the main text.

it is important not to conflate subjectivity with narcissism, for the subject is not only a subject to itself but is also a subject to and of others. Subjectivity is inherently dialogic, and it is in this sense that it has been at the heart of the novel since the inception of the genre in the eighteenth century. The novel is psychologically attuned to both interiority and to social context and in reading a novel, as Patricia Waugh suggests, as we 'hear more insistently and become aware of thought as a play of voice [...] we become aware too of how what is me is always already constituted out of the voices of the other'.[21] So in dramatizing interiority, the novel also engages with the intersubjective relations which are the starting point for a more objective understanding of the world. As Thomas Nagel has famously argued, subjectivity and objectivity are not opposed but exist on a continuum and we approach objectivity as we 'detach gradually from the contingencies of the self'.[22] The novel is centrally concerned with this continuum and with the convergence and divergence between subjective and objective points of view, and as Ian McEwan has suggested, it is uniquely able to mediate between these perspectives and register the points at which they are incommensurable.[23]

Robinson makes it clear that her objections to neo-Darwinism are related to her religious beliefs and in this respect her critique must be placed in the context of a wider pattern in the USA in which neo-Darwinian science was—and is—routinely pitted against creationism and intelligent design. This cultural schism forms the backdrop of Robinson's intervention and is implicit in Daniel Dennett's vigorous defence of neo-Darwinism in in *Darwin's Dangerous Idea*. Indeed, as John Dupré has noted, there is an isomorphic quality in the relations between religion and neo-Darwinism, as 'extreme neo-Darwinists sometimes share with creationists the yearning for an all-encompassing

---

[21] Patricia Waugh, 'The Novel as Therapy: Ministrations of Voice in an Age of Risk', *Journal of the British Academy* 3 (2015), 35–68 (50).
[22] Thomas Nagel, *The View from Nowhere* (Oxford: Oxford University Press, 1989), p. 5.
[23] Ian McEwan, 'Literature, Science, and Human Nature', in *The Literary Animal: Evolution and the Nature of Narrative*, ed. by Jonathan Gottschall and David Sloan Wilson (Evanston: Northwestern University Press, 2005), pp. 5–19 (p. 6).

scheme, a single explanatory framework that makes sense of life'.[24] However, in the largely post-religious British context there was little sense that neo-Darwinian theory posed a challenge to belief; it was more often perceived as a challenge to a residual humanism which was itself a substitute for religion. This is a point made by A.S. Byatt in an essay on post-war fiction in which she links F.R. Leavis's 'great tradition' of English literature with the Comtean 'Religion of Humanity' which had replaced religion in the nineteenth century. Writing in 1979, she suggests that the religion of humanity is being displaced in its turn and that that novelists now exist in an 'uneasy relation to the afterlife of these literary texts [. . .] They are the source of enlightenment, but not true.'[25] As we shall see, the fiction of Byatt and McEwan is marked by nostalgia for the kind of literary humanism invoked in Byatt's essay, which was under pressure not only from postmodernism and post-structuralism but from the scientific humanism espoused by figures like Wilson, Dawkins, and Pinker. Now more often known as secular humanism, scientific humanism rejects transcendentalism in all its forms and privileges scientific rationality as the means to knowledge.

An important hub for neo-Darwinism in the UK was the Darwin Seminar at the LSE which ran for over a decade from 1995 and was convened by Helena Cronin, a philosopher and key supporter of Dawkins.[26] The seminar, which was more of an intellectual salon, hosted influential speakers including John Maynard Smith, Daniel Dennett, and Steven Pinker and attracted audiences from across disciplines, including novelists such as Byatt and McEwan.[27] However, neo-Darwinism was generally viewed with hostility in the humanities and social sciences, due to the long history of the use of biology to justify oppression on the grounds of race and sexuality. Genetics is

[24] John Dupré, *Processes of Life: Essays in the Philosophy of Biology* (Oxford: Oxford University Press, 2012), p. 160. Dupré cites Daniel Dennett's *Darwin's Dangerous Idea* as an example of this tendency, as Dennett uses natural selection to explain 'everything from the breeding behaviour of bees to the deliberative processes of the human mind'.

[25] A.S. Byatt, 'People in Paper Houses: Attitudes to "Realism" and "Experiment" in English Post-war Fiction', in *Passions of the Mind: Selected Writings* (London: Vintage, 1993), p. 167.

[26] For example in the preface to the second edition of *The Selfish Gene* Dawkins warmly acknowledges Cronin's help with the new chapters in the book (p. xiii).

[27] An archive of the seminar is available at https://digital.library.lse.ac.uk/collections/publiclectures/subject under the heading 'Evolution (Biology)'.

implicated in this history by virtue of its association with eugenics, which Hilary Rose has identified as genetics' shadowy twin.[28] As the historians of science Staffan Müller-Wille and Hans-Jörg Rheinberger have shown, classical genetics developed in a context that was preoccupied with questions of eugenics, racial identity, and sexuality 'in short, a biopolitics of what came to be called the "racial body".'[29] After the Second World War, despite the revelation of Nazi atrocities, eminent geneticists including Francis Crick continued to endorse the idea that humanity should take control of its destiny through genetic manipulation of the population, as is evident in his comments at a 1962 CIBA symposium on 'Man and His Future'.[30] Neo-Darwinian theory is linked to eugenics through E.O. Wilson, who expressed support for both conventional eugenics and genetic engineering in his popular book *On Human Nature*. While no other neo-Darwinian expressed such views, Wilson's comments point to the continuing porosity of the border between genetic and eugenic thought. Neo-Darwinism also raised concerns because of its perceived genetic determinism, which could be used to underwrite racist and sexist stereotypes. To the extent that it argued for a universal human nature neo-Darwinism undermined the category of race, and in its guise as evolutionary psychology it explicitly rejected the race concept. Nonetheless, by continuing to explore the extent to which there might be genetic differences between races, figures like Wilson and Pinker retained the implicit connection between race and biological difference.[31] Neo-Darwinism is also predicated on the assumption that genetically inscribed differences between the sexes have evolved because of their differential investment in

[28] See Hilary Rose, 'Eugenics and Genetics: the Conjoint Twins?', *New Formations* 60 (Spring 2007), pp. 13–26 (p. 14).

[29] Staffan Müller-Wille and Hans Jörg Rheinberger, *A Cultural History of Heredity* (Chicago: Chicago University Press, 2012), p. 185.

[30] See Francis Crick, 'Discussion: Eugenics and Genetics', in *Man and His Future: A Ciba Foundation Volume*, ed. by Gordon Wolstenholme (Boston: Little, Brown & Co., 1963).

[31] For example, Steven Pinker notes that 'some racial distinctions [. . .] may have a degree of biological reality' in *The Blank Slate* (144). This was only a few years after Richard Herrnstein and Charles Murray were making the case for racial differences in intelligence in the infamous *The Bell Curve*. See Richard J. Herrnstein and Charles Murray, *The Bell Curve: Intelligence and Class Structure in American Life* (New York: Free Press, 1996).

reproduction, a perspective that lent itself to a conservative view of gender relations. It is therefore unsurprising that when Helena Cronin argued that 'evolution made men's and women's minds as unalike as it made their bodies' it drew this tart response from Deborah Cameron: 'I cannot help hearing the echoes of every misogynist thinker—Rousseau, Nietzsche, the fascists of the early twentieth century—who ever proclaimed the same doctrine [...] even dressed up in new Darwinian clothes, how can such concepts be compatible with feminism? The short answer is, they can't.'[32]

Throughout the period of neo-Darwinism's ascendency, then, the humanities and social sciences remained strongly resistant to biological explanations of human behaviour. A particular concern was genetic essentialism, that is, the belief that differences between sexes, races, and classes had a genetic basis and were to some degree innate. For second-wave feminism, driving a wedge between sex (as anatomical difference) and gender (as socially constructed identity) was a crucial political move, making it possible to contest existing gender roles and fashion a discourse of gender transformation. For scholars of race, it was equally important to distinguish between discredited biological conceptions of race and a view of race as a socially constructed category which was amenable to change.[33] As the dominant theory of knowledge in the social sciences and humanities in the late-twentieth century, social constructionism has been hugely important in revealing the extent to which identities are socially produced and imposed within hierarchies of power and privilege. However, as Shannon Sullivan has argued, the dichotomy between biology and constructionism is not only false but sets limits to our understanding of the effects of racism and sexism.[34] Her work is part of a current recalibration of the relationship between the humanities and the biological sciences which is impelled by the need to unravel this dichotomy and explore the mutual

[32] Deborah Cameron, 'Back to Nature', in *The Trouble & Strife Reader*, ed. by Deborah Cameron and Joan Scanlon (London: Bloomsbury, 2009), pp. 149–58 (p. 156).

[33] In a classic paper Richard Lewontin showed that human variation was overwhelmingly within rather than between races and argued that racial categories should be seen as socio-economic constructs. See Richard Lewontin,'The Apportionment of Human Diversity', *Evolutionary Biology* 6 (1972), 381–98.

[34] Shannon Sullivan, *The Physiology of Sexist and Racist Oppression* (New York: Oxford University Press, 2015), p. 5.

imbrication of the biological and the social. The anthropologists Jörg Niewohner and Margaret Lock can be seen as representative of this shift when they argue that it is no longer possible to dismiss biology as 'the enemy of critical thought'; rather, they suggest that the humanities and social sciences must engage with the life sciences in order to understand the materiality of human agency.[35]

As Sullivan points out, the rapprochement between the humanities and the biological sciences owes a great deal to the pioneering work of the feminist philosophers of biology Evelyn Fox Keller and Anne Fausto-Sterling. Fox Keller is a longstanding critic of gene-centric reductionism, who argues that the distance between the genotype (nature) and the phenotype (nurture) militates against any simple parsing of genetic causal factors. She rejects the attribution of agency to genes or to DNA, writing that 'by itself, DNA doesn't *do* anything' and insisting that DNA is 'embedded in an immensely complex and entangled system of interacting resources that are, collectively, what give rise to the development of traits'.[36] She stresses the openness of the genome to environmental factors within the cellular, internal, and external environment, and suggests that in consequence it makes no sense to separate developmental influences into the categories of the genetic and the environmental. There are many points of contact between her work and that of Fausto-Sterling, which focuses on the biology of gender, integrating biological and social perspectives in its account of development. Just as Fox Keller argues that there is no space between nature and nurture, Fausto-Sterling contends that 'we are always 100 per cent nature and 100 per cent nurture'.[37] Like Fox Keller, she argues that bodies are unbounded and contends that to understand sex and gender identifications we need to examine how experiences of pleasure, aversion, and desire become embodied. Moreover, she suggests that racial and class discrimination can become embodied as we 'physically imbibe' experiences such as stress and physical trauma. With Fox Keller, she has been a crucial interlocutor for scholars in

---

[35] Jörg Niewöhner and Margaret Lock, 'Situating Local Biologies: Anthropological Perspectives on Environment/Human Entanglements', *BioSocieties* 13, 681–97 (p. 693).

[36] Evelyn Fox Keller, *The Mirage of a Space between Nature and Nurture* (Durham and London: Duke University Press, 2010), p. 51.

[37] Anne Fausto-Sterling, 'The Bare Bones of Sex: Part 1 – Sex and Gender', *Signs* 30 (2005), 1491–527 (1510).

the humanities and social sciences seeking to tease out the philosophical implications of what Diana Coole and Samantha Frost term 'the
new biology'.[38] Building on this perspective, the feminist philosopher
Rosi Braidotti has urged humanities scholars to theorize 'via and with
science', an approach exemplified in her articulation of a 'matter-realist'
philosophy that takes its inspiration from postgenomics and from the
autopoietic theories of Maturana and Varela (158). Her philosophy is
predicated on the idea that matter is self-organizing and that subjectivity
should be seen in terms of autopoiesis. In addition, she stresses that
the embodied subject is relational and caught up in complex co-
dependencies with non-human organisms and with social, psychic, and
ecological environments. This relational perspective suggests the need for
a 'transversal ethic' which would be grounded in the interdependency of
human and non-human actors and would redefine kinship and ethical
accountability so as to 'rethink links of affectivity and responsibility [...]
for non-anthropomorphic organic others' (103). Braidotti aims to contest what she sees as a melancholic tendency in the work of contemporary thinkers such as Judith Butler and Paul Gilroy, arguing that her
brand of vital materialism can help to actualize 'new concepts, affects and
planetary subject formations' organized around *zoe* as a generative life-
force (104). Such an affirmative philosophy could also be mobilized to
challenge the bio-genetic structure of contemporary capitalism which
profits precisely from the commodification of life itself.[39]

However, it is Catherine Malabou whose work represents the most
ambitious attempt to build bridges between philosophy and the life
sciences. In her early book *What Should We Do with Our Brain?* a
concept of plasticity initially derived from Hegel is mapped onto the
biological concept of neuroplasticity, that is, the continuous formation
and reformation of synaptic connections in the brain. Drawing on
the work of the neuroscientist Joseph LeDoux, Malabou argues that
the extent of genetic determination is limited and that as a form of
epigenetic sculpting, neuroplasticity offers a model of subjectivity in

---

[38] *New Materialisms: Ontology, Agency and Politics*, ed. by Diana Coole and Samantha
Frost (Durham and London: Duke University Press, 2010), p. 16.
[39] For an extended discussion of this point see Nikolas Rose, *The Politics of Life Itself:
Biomedicine, Power, and Subjectivity in the Twenty-First Century* (Princeton and Oxford:
Princeton University Press, 2007).

terms of continuous self-fashioning. This, in turn, has significant implications for our understanding of human agency, because genetic non-determination opens 'the possibility of a social and political non-determinism, in a word, a new freedom'.[40] Malabou pursues this line of thought in subsequent books including *Plasticity at the Dusk of Writing* and *The New Wounded* but it is in *Before Tomorrow: Epigenesis and Rationality* that she develops her most complex and sustained account of the relationship between the biological and the logical, the natural and the transcendental. Taking Kant's discussion of the epigenesis of reason as her starting point, she argues that the a priori categories of thought are realized at the point of contact between the categories and experience and that they exist only in their materialization, in their encounter with life. In this respect they resemble genes, which from an epigenetic point of view are inert and meaningless until they are activated or expressed. Co-articulating biological and philosophical perspectives, Malabou suggests that the a priori categories of thought must be understood as, in Ian James' words, 'folded into the temporal and material becoming of epigenetic development'.[41] Moreover, because 'the a priori has no meaning' reason has no fixed origin or ground: rationality engenders itself out of this necessary lack.[42] Both the transcendence of thought and the substance of biological form rest on an ontological void and it is this shared groundlessness which enables continuity and exchange between the biological and the logical. An emphasis on change and exchangeability also informs Malabou's account of the implications of research in cloning, which has shown that cellular differentiation can be reversed and that adult cells can be de-differentiated or reprogrammed. As Malabou points out, both processes represent the recovery of properties which it was thought had been lost, so that at the heart of contemporary biological research lies the reactivation of 'ancient forms of life'; this in turn suggests to

[40] Catherine Malabou, *What Should We Do with Our Brain?*, trans. by Sebastian Rand (New York: Fordham University Press, 2008), p. 13.

[41] Ian James, '(N)europlasticity, Epigenesis and the Void', *parrhesia* 25 (2016), 1–19 (11).

[42] Catherine Malabou, *Before Tomorrow: Epigenesis and Rationality*, trans. by Carolyn Shread (Cambridge: Polity Press, 2016), p. 98.

Malabou that there are potentials within the living that have not yet been exhausted by evolutionary history.[43]

Epigenetics is also of major interest to the social sciences, as epigenetic mutations are seen as responses to the psychosocial as well as the physical environment, a means by which experience gets under the skin. While research in this area is in its early stages, there is evidence that 'social insult' and the experience of trauma and violence are epigenetically mediated and that such epigenetic changes can be reproduced across generations. For sociologists like Paul Martin, the hope is that opening the black box of biology in this way will enable a fuller understanding of the somatic effects of inequality and of class-based, gendered, and racial hierarchies of oppression. As yet it is unclear how such insights will come to inform public policy, but Martin takes a positive view of epigenetics' potential to shape a progressive social agenda.[44] Maurizio Meloni is more pessimistic, arguing that the return to neo-Lamarckian ideas of the inheritance of experience could lead to a stigmatizing of groups considered to be at risk. As an example, he cites a study of the correlation between socio-economic status and levels of DNA methylation in Glasgow. The study found that DNA hypomethylation was associated with the most deprived group of participants and in consequence aberrant methylation was proposed as a biomarker of social adversity, neglect, and poverty.[45] As Meloni suggests, the danger is that once such groups are identified as being at risk, they could become the target of a new epigenetic biopolitics. The emancipatory potential of epigenetics, which lies in its focus on environments which we have the power to change, could be lost if an emphasis on the indelible scars of the past leads to a view of the disadvantaged as somehow biologically distinct.[46] At the other end of

[43] Catherine Malabou, 'One Life Only: Biological Resistance, Political Resistance', trans. by Carolyn Shread, *Critical Inquiry* 42 (2016), 429–38 (available at https://criticalinquiry.uchicago.edu/one_life_only/> [accessed 12 Apr 2019].

[44] Paul Martin, 'Toward a Biology of Social Experience?', response to Margaret Lock, 'Comprehending the Body in the Era of the Epigenome', *Current Anthropology* 56, 151–277 (167–8) (April 2015).

[45] McGuinness, D. et al., 'Socio-economic status is associated with epigenetic differences in the pSoBid cohort', *International Journal of Epidemiology* 41 (9 Jan. 2012), 151–60. (doi:10.1093/ije/dyr215).

[46] Maurizio Meloni, response to Margaret Lock, 'Comprehending the Body in the Era of the Epigenome', *Current Anthropology* 56, 151–277 (168) (April 2015). In the UK,

the social scale, it is easy to see how the concept of the malleable genome could sponsor personalized regimes aimed at training and epigenetically modifying one's genome. Such Foucauldian care of the self would not, of course, be available to those without the requisite economic and social capital.

As these interdisciplinary engagements suggest, postgenomic science has become as culturally pervasive as the genetic discourse which it has to a significant extent displaced. From the outset genetics has been entangled with meanings which extend far beyond the biological but by the late-twentieth century, the gene had come to represent, in Judith Roof's words, 'a cosmic truth, representative of all life, residence of all answers, potential for all cures, repository for all identity, end to all stories'.[47] Its iconic status was encouraged by the fact that exponents of neo-Darwinism depicted the action of genes in anthropomorphic terms, enabling a transference of agency from humans to genes in the reader's imagination. Moreover, their focus on Darwinian fitness led to explanations of human behaviour which privileged competitive struggle and for this reason, critics have detected a link between the rise of neo-Darwinism and the dominance of neoliberal ideology as promoted by Margaret Thatcher and Ronald Reagan. The primatologist Frans de Waal, for example, has argued that *The Selfish Gene* endorsed Thatcherite ideology by teaching that 'since evolution helps those who help themselves, selfishness should be looked at as a driving force for change rather than a flaw that drags us down'.[48] On one level this is to misread Dawkins' text and take literally metaphors and inset narratives that are intended to illustrate genetic strategies; moreover in a footnote to the second edition of *The Selfish Gene* Dawkins explicitly distances himself from the 'new right' Conservative government that came to power in 1979 and which 'elevated meanness and selfishness to the status of ideology' (268) On another level, it can be argued that there is a homology between neo-Darwinism and neoliberalism in that both

---

there is a long history of thinking about the poor in terms of a biologically distinct underclass, from the 'social residuum' of the 1880s to the post-war 'problem family'.

[47] Judith Roof, *The Poetics of DNA* (Minneapolis: University of Minnesota Press, 2007), p. 2.

[48] Frans de Waal, *Our Inner Ape: The Best and Worst of Human Nature* (London: Granta Books, 2006), p. 21.

postulate external laws which govern the success or failure of atomistic entities competing for resources: in this sense, there is indeed a coherence between neo-Darwinian natural selection and the logic of the free market.

With the shift to postgenomics, the question of agency has been radically reconfigured. If genes are modified by experience, our fate is not predetermined and in the words of the epidemiologist Tim Spector, we can change both our genes and our destiny.[49] Invoking the idea of transgenerational epigenetic inheritance, Spector points out that in modifying your genes through lifestyle you can promote the health of your children and grandchildren, while it is also the case that genes can be adversely affected by environmental factors outside individual control, with the damage being passed down the generations. Accordingly, there is a tension in the epigenetic imaginary between the promise of positive agentic change and the risk of involuntary exposure, and this in turn feeds into debates about individual and collective responsibility. At the level of the individual we are encouraged to consider the impact of our choices not just on our own health but that of future generations, while at the social level there is growing pressure to hold state actors and corporations accountable for the release of pollutants and other environmental toxins. In this respect, there is an intersection between postgenomic and post-anthropocentric thought, for as Margaret Lock has pointed out, discussion of the effects of climate change must now factor in the impact on human health of environmental toxins which may be epigenetically inscribed.[50] Research in epigenetics is thus contributing to an understanding of humans and the environment as mutually constitutive, a perspective far from the neo-Darwinian view of humans as bounded and agentic individuals.

This book maps the arc of literature's engagement with these changes in biological thought. The first three chapters explore responses to neo-Darwinism which range from sharp critique to convergence, sometimes within the same text. In Chapter 1 Doris Lessing's science fiction sequence *Canopus in Argos* is read as a critical response to the

---

[49] Tim Spector, *Identically Different: Why You Can Change Your Genes* (London: Weidenfeld & Nicholson, 2012), p. 293.

[50] Margaret Lock, 'Mutable Environments and Permeable Human Bodies', *Journal of the Royal Anthropological Institute (NS)*, 24 (September 2018,) 449–74 (p. 455).

sociobiology controversy of the 1970s. In a radical move, Lessing mobilizes the Sufi theory of 'cosmic evolution' to counter the atomism and reductionism of sociobiology, stressing the coevolution of physical and spiritual life and critiquing a scientific rationality which takes no account of the subjectivity of the observer. In this respect, there is a suggestive convergence between her fiction and A.N. Whitehead's critique of scientific materialism. As Lessing's protagonists reflect on their efforts to comprehend the material universe, their insights resonate with Whitehead's attempt to overcome what he called the 'bifurcation of nature', that is the division between 'the nature apprehended in awareness and the nature which is the cause of awareness'.[51] For Lessing as for Whitehead, our immediate sensory perceptions and the hidden molecular world have an equal ontological status. In addition to pushing at the limits of reductionism, Lessing challenges the biologizing of stereotypes of sex and race which is a corollary of sociobiology's genetic determinism. *The Marriages between Zones Three, Four and Five* undertakes a patient deconstruction of sociobiological claims for innate sexual difference, exploring the epigenetic dynamics of development and excavating the biosocial formation of female desire and maternal attachment. And the sequence as a whole can be read as a response to the latent racism of sociobiology and as a riposte to E.O. Wilson's speculations about a genetically engineered future. In Lessing's intergalactic empires, the category of race is shown to be central to expansion and wealth extraction, as 'degenerate' populations are subjected to eugenic and genetic manipulation to improve efficiency. Moreover, as Lessing suggests through specific allusions to the British imperial past, the intergalactic future mirrors the history of Western colonialism, for which biological racism is a structuring principle.

Chapter 2 turns to A.S. Byatt, placing the four-part novel sequence known as the *Quartet* in the context of Foucault's *The Order of Things*, where he analyses the emergence of biology as a style of thought and announces the death of Man as the ruling category of the modern *episteme*. Writing two decades on from Foucault, Byatt is less sceptical of the ontological claims of biology and is fascinated by neo-Darwinian

---

[51] Alfred North Whitehead, *The Concept of Nature* (Cambridge: Cambridge University Press, 1920), pp. 30–1.

explanations of human behaviour. A central issue is the innativism–constructionism opposition, which she broaches with reference to a 1975 debate between Noam Chomsky and Jean Piaget in which Chomsky championed the 'crystal' model of living beings, predicated on the existence of a determining genetic programme, while Piaget endorsed the 'flame' model associated with the autopoietic principle of order from noise. As Byatt's protagonists, including geneticists, linguists, and philosophers, explore the implications of these models, the sequence builds towards an endorsement of the neo-Darwinian view of human behaviour as governed by a genetic core which is impervious to culture. However, this view is modified in Byatt's representation of the sexual and reproductive life of the main protagonist, Frederica Potter. Taking up the question posed by John Maynard Smith, of why sex evolved in the first place, Byatt contrasts the elegant simplicity of parthenogenesis with the conflict generated by sexual reproduction and concludes that while it may enhance genetic variation, the costs of sex are high for women. Frederica's experience seems to bear out the neo-Darwinian axiom that women have a greater biological investment in reproduction and for this reason take greater long-term responsibility for children.[52] However, Byatt suggests that changes in women's lives, particularly in relation to education and employment, are set to challenge some of the assumptions about women's 'nature' which underpin the work of evolutionary biologists such as George C. Williams.

The third chapter explores Ian McEwan's interest in neo-Darwinian theory in its manifestation as evolutionary psychology. Exponents of evolutionary psychology take the gene's eye view popularized by Richard Dawkins and argue that human behaviour is driven by 'mental modules' which have been forged by genes to ensure their own proliferation. In their view, society is a network delicately balanced between the twin poles of competition and cooperation, both of which ultimately serve genetic self-interest. In *Enduring Love*, McEwan assesses the implications of this for our understanding of personal relations,

---

[52] Anne Fausto-Sterling has been highly critical of such accounts of sexual difference, pointing out for example that there is no evidence to support the claim that sperm production is less costly than egg production. See Anne Fausto-Sterling, 'Beyond Difference: Feminism and Evolutionary Psychology', in *Alas, Poor Darwin: Arguments Against Evolutionary Psychology*, ed. by Hilary Rose and Steven Rose (London: Vintage, 2001), pp. 174–89 (p. 176).

religion, and the arts and considers the extent to which his protagonists display the tendency towards self-deception which is a central theme in evolutionary psychology. This theme is pursued in *Atonement*, which explores the self-deception of a character who turns out to be the author of the novel we are reading, the implication being that literary realism may be impossible as the experiences on which writers draw are 'warped by a prism of desire and belief'.[53] *Atonement* also dramatizes the tension between the universal and the specific in human nature, a theme which is addressed in *Saturday* as McEwan explores the human universal of warfare in the context of the run-up to the Iraq war. The novel turns on the analogy between conflicts in the protagonists' personal lives and those on the geopolitical stage. In both cases, male dominance hierarchies and self-deception are in play, as McEwan suggests that the actions of political leaders (as well as those of middle-class professionals) are shaped by genetically mediated dispositions of which they may be unaware. The novel shares many concerns with Steven Pinker's *The Blank Slate* and reflects the neo-conservative politics associated with evolutionary psychology. However, having argued that neo-Darwinian explanatory narratives are the armature of much of McEwan's fiction, the chapter concludes by reading his 2016 novel *Nutshell* as a witty recantation of these ideas, as the foetus-narrator casts his development in terms of autopoietic self-fashioning and explicitly rejects the concept of genetic self-interest.

Chapters 4 and 5 are concerned with literary responses to the new biological perspectives that were emerging around the turn of the millennium. Chapter 4 focuses on cloning, which came to public attention with the publicity surrounding the birth of Dolly the sheep in 1996. The chapter opens with Eva Hoffman's *The Secret*, in which the clone stands as a metaphor for second-generation Holocaust survivors haunted by the sense that they lack authenticity as their experience cannot measure up to their parents' ordeals. Hoffman also explores the potential of cloning to dislodge sexual reproduction as the cornerstone of the social order and to queer structures of kinship, while its implications for our understanding of living beings are examined through the consciousness of the clone-narrator, who comes to understand identity

---

[53] Ian McEwan, *Enduring Love* (London: Vintage, 2006), p. 180.

in terms of recursive self-fashioning rather than genetic replication. Kazuo Ishiguro's *Never Let Me Go* also invokes the legacy of the Holocaust to suggest the continuities between twentieth-century bio-politics and the biopolitical order of the twenty-first century. The clones in this novel are read in the light of Giorgio Agamben's theorization of bare life and his analysis of the state of exception which allows for the detention of those who have been relegated to this category, notably refugees and asylum seekers. Ishiguro's clones resemble refugees in that they are excluded from the protection of the juridical order, are segregated from the wider population, and are constructed as racial others and/or as animals. However, in negotiating their exclusion and exposure to death, the clones develop a way of being in the world which is attuned to the potential of what Agamben calls 'indifference', or the pre-relational zone from which differences emerge. In their relationships with each other, they attend to the singularity of each other's being in a way which anticipates Agamben's concept of a 'coming community' which is not based on humanist ideas of personhood.[54]

The fifth chapter explores literary texts which develop postgenomic themes. Margaret Drabble's *The Peppered Moth* was published before epigenetics began to impinge on public consciousness but invokes the precursor discourse of neo-Lamarckism to challenge the neo-Darwinian view of inheritance. Focusing on a working-class girl's experience of social defeat and depression in the early twentieth century, it underscores the somatic impact of such experience and suggests that it reverberates across the generations, cohering with recent research which has found that social defeat is associated with forms of depression which are epigenetically mediated. Anticipating current controversies, the novel draws attention to the possibility that disadvantage could be biologically transmitted, leading to the stigmatization of particular social groups and a form of epigenetic determinism. Jackie Kay's memoir *Red Dust Road* also starts with neo-Darwinian theories of inheritance but pushes against the limits of this explanatory framework in its representation of her adoption and tracing of her birth parents. In this narrative of origins, nurture as well as nature is shown to be somatically inscribed, as the love and care of Kay's adoptive

---

[54] See Giorgio Agamben, *The Coming Community*, trans. Michael Hardt (Minneapolis: University of Minnesota Press, 1993).

mother enables her to thrive against the odds in the early months of life. This engenders a resilience which supports Kay when she has a life-changing accident at the age of sixteen: in mapping the way she reconfigures her identity in the wake of the accident, *Red Dust Road* echoes Catherine Malabou's celebration of the plasticity and resilience of the living being. In contrast, the negative dimension of plasticity is apparent in the story of Kay's brother Maxie, whose experience of racism and rejection becomes a biological inheritance, in line with findings which suggest that the effects of racism can be epigenetically inscribed. Emphasizing the porosity and unboundedness of the body, Kay's memoir points to the need for an expanded biological imaginary which recognizes the foundational nature of our interdependency with our environments.

1

Doris Lessing's Evolutionary Epic

For Doris Lessing, writing in 1978, it was self-evident that science fiction was the literary form which had most effectively captured the impact of twentieth-century science. In her Preface to *Shikasta*, she points out that the genre has both reflected and anticipated the 'new worlds' that science has opened up, mapping them in dazzling fashion.[1] In this spirit, this chapter reads Lessing's science fiction as a reflection on the sociobiological project of explaining human nature and an anticipation of the potential implications of molecular neo-Darwinism. Lessing's interest in this science—and in the science fiction genre—is already clear in *The Four-Gated City* (1969), the final volume of the 'Children of Violence' sequence, which concludes with a vision of an evolutionary leap brought about by atomic radiation. Nonetheless, her sustained engagement with science fiction in the 1980s surprised many readers and proved controversial. Ursula K Le Guin, for example, was critical of the first volume of what Lessing called her 'space fiction' series, noting an unevenness of tone which she ascribed to Lessing's lack of interest in alien worlds and her tendency to didacticism.[2] For Le

---

[1] Doris Lessing, *Canopus in Argos: Archives Re: Colonised Planet 5 Shikasta* (St Albans: Granada Publishing, 1981), 'Preface', np. Further page references will be given within the main text.

[2] Lessing's rejection of the term science fiction may have been linked to the fact that it was (and arguably still is) considered a lowly genre. Her move can be compared with that of Margaret Atwood who rejects the term science fiction for her work, preferring

Genetics and the Literary Imagination. Clare Hanson, Oxford University Press (2020). © Clare Hanson.
DOI: 10.1093/oso/9780198813286.001.0001

Guin, *Shikasta* was redeemed only by its 'exact, brilliant, compassionate' portrayal of individuals, that is, by the strand of realism which runs through it.[3] This chapter suggests that the novel's disjointed quality and abrupt shifts of register are inseparable from its cross-disciplinary scope and ambition. Together with the other *Canopus in Argos* novels, *Shikasta* constitutes a significant response to the rise of popular science in the late twentieth century and its incursion into the traditional territory of the humanities. Lessing engages with the sociobiological arguments of Edward O. Wilson and Richard Dawkins as they too move across disciplines, extrapolating from mid twentieth-century genetics to make broader claims about human nature. This chapter explores the richness of Lessing's reflections on these ideas in the Canopus novels, which are too often dismissed as a failed experiment in the science fiction genre.[4]

One of the most striking features of Wilson's *Sociobiology* (1975), the book which first mapped out the sociobiological project, is its contention that the extension of evolutionary biology which it proposes has an explanatory power which surpasses that of all other disciplines which focus on human nature. In the chapter which deals with man, Wilson presents himself as a 'zoologist from another planet', shaping a world history which will draw not only on genetics, zoology, and ethology but also on anthropology, sociology, and history. This suggests a laudable degree of interdisciplinarity but rather than granting equal status to these disciplines, Wilson goes on to claim that from a macroscopic point of view, 'the humanities and social sciences shrink to specialized branches of biology; history, biography and fiction are the research protocols of human ethology'.[5] In similar vein, Richard Dawkins opens

'speculative fiction'. See Margaret Atwood, '*The Handmaid's Tale* and *Oryx and Crake* in context', *PMLA* 119 (May 2004), 513–7 (513). doi: 10.1632/003081204X20578.

[3] Ursula K. Le Guin, 'Doris Lessing's First Sci-Fi Novel Reads Like a Debut Novel', *New Republic* 13 Oct. 1979 https://newrepublic.com/article/115631/doris-lessing-shikasta-reviewed-ursula-le-guin [accessed 25 May 2017].

[4] The interpretation in this chapter owes a great deal to conversations at *Doris Lessing 2014: An International Conference*, held at Plymouth University in September 2014. I am especially grateful to Lara Choksey and David Sergeant, who persuaded me that in the *Canopus* sequence Lessing orchestrates a more complex response to genetics than I had thought.

[5] Edward O. Wilson, *Sociobiology: The New Synthesis* (Cambridge, MA: Harvard University Press, 1975), p. 547. Further page references will be given within the main text.

The Selfish Gene by lamenting the fact that 'philosophy and the subjects known as "humanities" are still taught almost as though Darwin had never lived' before going on to emphasize the importance of evolutionary biology for understanding 'every aspect of our social lives, our loving and hating, fighting and cooperating, giving and stealing'.[6] Wilson and Dawkins differ on many issues but both are convinced of the power of biology to explain what Wilson calls 'man's ultimate nature'.[7] Together, they put the biological basis of human behaviour firmly back on the agenda at a time when constructionist explanations predominated in the social sciences, particularly in relation to issues of race and gender. In so doing, they provoked heated debate among academics and members of the wider public.

A key point of contention was the genetic determinism which was widely perceived to be at the heart of the sociobiological enterprise and which was energetically attacked. As Philip Kitcher has argued, the charge of genetic determinism was unfair in that Wilson and his followers acknowledged that the development of the phenotype was dependent on both genes and the environment.[8] However, lurking beneath debates about determinism in which each side caricatured the other, there lay a genuine disagreement. As Kitcher puts it:

> Wilson and his followers believe that the values of the [biological] functions vary relatively little and that they do only when the environment is quite drastically altered. The critics maintain that the values of the functions are quite responsive to changes in environmental variables. Each side may justly claim to have absorbed the commonplace story about genes and development. There is also a genuine difference, deriving from alterative articulations of the story.   (25)

---

[6] See Richard Dawkins, *The Selfish Gene*, 2nd edn (Oxford: Oxford University Press, 1989, pp. 1–2. Further page references will be given within the main text. See also Richard Dawkins, *The Extended Phenotype*, with an afterword by Daniel Dennett (Oxford: Oxford University Press, 1999), p. 265). Further page references will be given within the main text. Dennett praises *The Extended Phenotype* as a contribution which spans the disciplines of science and philosophy.

[7] Edward O. Wilson, *On Human Nature*, with a new preface (Cambridge, MA: Harvard University Press, 2004), p. 1. Further page references will be given within the main text.

[8] Philip Kitcher, *Vaulting Ambition: Sociobiology and the Quest for Human Nature* (Cambridge, MA: MIT Press, 1987), p. 25.

What is at stake is the extent to which the functions set by the genotype are fixed and unchanging, sealed off from experience and the environment. For Wilson and his followers they are relatively fixed, while critics such as Stephen Jay Gould see them as more plastic and open to environmental influences. As Kitcher's comments indicate, the competing perspectives arise from different interpretations of the evidence and are articulated as alternative versions of the evolutionary narrative. Further, to make their point of view accessible, both sides use literary techniques such as metaphor and intertextual allusion. In *Sociobiology*, for example, while acknowledging that culture is largely independent of genes, E. O. Wilson argues that human beings have a 'drive towards total knowledge, right down to the levels of the neuron and the gene'. He then suggests that once we can explain ourselves in these terms, the result may be 'hard to accept' and closes his discussion with a quotation from Camus's 'Myth of Sisyphus' which represents man (*sic*) as a stranger living 'in a universe divested of illusions and lights' (575). Somewhat paradoxically, he deploys cultural resources to argue that genetic explanations operate at a level which is both deeper than culture and in opposition to it.

In his Pulitzer prize-winning sequel *On Human Nature* Wilson appears to reaffirm this point when he identifies a tension between cultural evolution, which aspires to higher ethical values, and genetic evolution. Posing the question of whether cultural evolution can ever completely replace genetic evolution, he argues that it is unlikely because 'the genes hold culture on a leash' and human values are 'constrained according to their effects on the gene pool'. For Wilson, all human behaviour, together with the 'deepest capacities for emotional response which drive and guide it', is no more than a 'circuitous technique' for keeping genetic material intact: ethics and morality have 'no other demonstrable ultimate function' (167). Again, whatever Wilson's intention, the choice of words (leash, constraint, ultimate function) implies that genes play a major part in shaping human behaviour, leading the reader to wonder about the precise status of cultural evolution. He goes on to argue that armed with 'the genetic rules of human nature', we should work towards a society structured in accordance with this knowledge. This process could be supported by 'conventional eugenics', in other words by manipulation of the gene pool at the population level, while in the longer term, genetic engineering and selection through cloning could enable directed evolutionary change (208).[9] Wilson does not explicitly endorse eugenics but nor does he repudiate eugenic strategies, which is startling given that he is writing at a time of continued tension over race in the post-Civil Rights period in the USA. He sees himself as a political liberal but his writing is scattered with statements that could be—and were—taken to suggest that distinctions of race and gender stemmed from genetic biases. He was taken aback by the controversy aroused by his claims about human nature and in his 1995 autobiography *Naturalist* attempted a more nuanced formulation of his understanding of genetic determinism, writing that 'although people have free will and the choice to turn in many directions, the channels of their psychological development are nevertheless—however much we might wish otherwise—cut more deeply by the genes in certain directions than in others'.[10] Clearly this later statement allows for a degree of human agency, but the nature of this agency is unclear and as Le Guin notes in a tart review of the book, Wilson's 'flat statement of belief that "people have free will"' is less than persuasive.[11]

The charge of genetic determinism was also levelled at *The Selfish Gene* and Dawkins responded to it in the second edition of the book, arguing that genes determine behaviour only in a statistical sense and that 'there is no reason why the influence of genes cannot easily be reversed by other influences' (268).[12] In *The Extended Phenotype*, he devotes an entire chapter to the issue, explaining that 'genetic variance

[9] The eugenic implications of Wilson's arguments are explored by the left-wing biologists Steven Rose, Richard Lewontin, and Leon Kamin in their well-known polemic against sociobiology *Not In Our Genes*. As they also point out, the reductionism and genetic determinism associated with sociobiology could be mobilized to legitimate existing social inequalities and to underwrite the shift away from the welfare state that was triggered by the economic crises of the 1970s.

[10] Edward O. Wilson, *Naturalist* (Washington, DC: Island Press, 1994), pp. 332–3.

[11] Ursula K. Le Guin, *The Wave in the Mind: Talks and Essays on the Writer, the Reader and the Imagination* (Boston, MA: Shambhala Publications Inc., 2004), pp. 159. Le Guin attacks Wilson's arguments in *Naturalist* on the grounds that his definitions are loose and he speculates beyond what is warranted by the evidence. In relation to his claims for the genetic determination of ethical behaviour she writes that 'we have a right to ask anybody who asserts that there are universal human moralities to list and define them. If he asserts that they are genetically determined, he should be able to specify the genetic mechanism and the evolutionary advantage they involve.'

[12] Dawkins makes these points in a footnote to the second edition which supplements his argument in Chapter 1 that 'our genes may instruct us to be selfish, but we are not necessarily compelled to obey them all our lives' (p. 3).

is a significant cause of much phenotypic variance in observed populations, but its effects may be overridden, modified, enhanced or reversed by other causes. Genes may modify the effects of other genes, and may modify the effects of the environment' (13). As with Wilson, the charge of genetic determinism is unjust in that Dawkins subscribes to the concept of gene–environment interaction and is clear that the effects of genes are not irrevocable. However, *The Selfish Gene* gives a different *impression* precisely because of the rhetorical skills that ensured the book's success, namely the use of metaphor and analogy. The titular metaphor, for example, conjoins and confuses the spheres of conscious and unconscious life and as it is developed through the conceit of genes as replicators and their human vehicles as 'gigantic lumbering robots', the characteristics of the human and the mechanical, the intentional and the robotic, are transposed to uncanny and unnerving effect. The use of extended analogy creates further ambiguities. The aim of the central section of the book is to explain the evolutionary game theory which was developed by W.D. Hamilton and Maynard Smith to model gene frequencies within populations. This is a mathematical theory concerned with risks and probabilities, but because Dawkins illustrates it with copious examples drawn from animal behaviour, the effect—as with the deployment of the metaphor of the selfish gene—is to confuse the distinction between genes and organisms. In a discussion of Maynard Smith's concept of the evolutionarily stable strategy (ESS) for example, detailed descriptions of 'hawk' and 'dove' behaviour, of strategists named 'retaliator' and 'bully' and of 'wars of attrition' serve only to reinforce the impression that animal and human behaviour is driven by selfish genes. It was not just lay readers who gained this impression: when the evolutionary psychologist Randolph M. Nesse first read the book he found it extremely disturbing, as 'much altruism of which I was personally proud was suddenly reframed as just another way my genes get me to do what benefits them. Selfish robots lumbered about in my dreams' (204).[13]

---

[13] Randolph M. Nesse, 'Why a Lot of People with Selfish Genes are Pretty Nice Except for their Hatred of *The Selfish Gene*', in *Richard Dawkins: How a Scientist Changed the Way We Think*, edited by Alan Grafen and Mark Ridley (Oxford: Oxford University Press, 2006), p. 204.

As this suggests, many of the debates around sociobiology turn on questions of interpretation which are linked to its exponents' appropriation of literary techniques for expository purposes. By using metaphor, extended analogy, and inset narrative, these writers introduce multivalence and indeterminacy into their texts and open them up to multiple interpretations. Dawkins' case is particularly striking as he not only uses metaphor freely but also deploys narrative perspective in a distinctive way. In *The Selfish Gene* the 'gene's eye view' of evolution becomes the narrative point of view and the reader is invited to identify with the agenda of the genes, rather as in an adventure story. In addition, Dawkins uses what he calls 'subjective soliloquy', in other words, interior monologue, to illustrate the calculations that underlie animal behaviour. For example, introducing the concept of the ESS he adopts the perspective of an elephant seal as he dramatizes the kind of (unconscious) reasoning that might precede an animal's decision whether or not to fight: 'He holds a valuable resource, which is why I want to fight him... But why does he hold it? Perhaps he won it in combat. He has probably beaten off other challengers before me' (68–9). While these strategies are undoubtedly effective, they can also operate as camouflage, obscuring aspects of the scientific argument which are potentially problematic. For example, Dawkins' representation of the gene as agentic masks an underlying problem in genetic thought, the difficulty of knowing whether the gene is the subject or object of 'gene action'.[14]

As we have seen, Wilson's literary techniques include metaphor (as in his invocation of a genetic leash), together with inset narratives drawn from anthropology and ethology. In the chapter on aggression in *On Human Nature* for example, he draws on the work of the anthropologist Napoleon Chagnon to offer a vivid account of life among the indigenous Yanomamö of Venezuela, arguing that the Yanamamö conduct their wars over women and that although a

---

[14] This is a point which is explored in Evelyn Fox Keller's *Making Sense of Life: Explaining Biological Development with Models, Metaphors and Machines* (Cambridge, MA: Harvard University Press, 2002). In a discussion of the ambivalence of the concept of the gene she writes that the gene is 'defined as an entity embodying the capacity to act (in whatever ways are required) in its own being'. However, it is simultaneously likened 'to a chemical object (especially an enzyme)' (p. 130). As Keller also suggests, it is the ambivalence of this concept which has made it so productive in scientific terms (p. 147).

quarter of the men die in battle, those who survive are 'wildly successful in the game of reproduction' (115). Wilson's example is unfortunate, as Chagnon's work is now seen as making an ungrounded claim for belligerence as a principle of social evolution. Despite being clear that 'the learning rules of violent aggression are now obsolete', Wilson follows Chagnon to the extent that he argues that in humans there is 'a genetic predisposition to learn aggressive responses' (114). In addition to anthropology, Wilson draws on the language of religion to add weight to the sociobiological case, transferring its cultural authority to his evolutionary narrative. He argues that science has probed 'distances and mysteries beyond the imagination of earlier generations' and it has even met God's challenge in the Book of Job (38:18): 'have you comprehended the vast expanse of the world? Come tell me all this, if you know'—Wilson's bold answer is that 'we *do* know, and we have told' (202). He adopts a similar tactic in relation to literature: his writings are studded with allusions to writers from John Bunyan to James Joyce, and in particular Camus. In the opening pages of *Sociobiology* he aligns Camus's view of the universe's indifference to suffering with the evolutionary view that human existence is designed not for happiness but 'favour[ing] the maximum transmission of the controlling genes' (4).

The fact that Wilson's and Dawkins' books were bestsellers was due in part to the authors' rhetorical skills but Wilson yearned for something more and in *On Human Nature* he called on scientists to work with the humanities to forge an 'evolutionary epic' that would convey the grandeur of evolutionary biology (201). However, his understanding of the collaborative process is in terms of a one-way transmission, whereby those great writers 'who can trouble and move the deeper reaches of the mind' are enlisted to convey the meaning of a science with which they must engage 'on its own terms' (203). Lessing could be said to have responded to this call but not as a fluent spokesperson for scientific orthodoxy, far from it: her alternative evolutionary epic stages a number of thought experiments which test the limits of sociobiology's explanatory power. It is likely that her interest in this field was stimulated by her friendship with Naomi Mitchison, whose brother was the eminent geneticist J.B.S. Haldane and whose circle included other key figures in the field such as Joseph Needham, C.H. Waddington and James Watson. Lessing comments in her autobiography on the 'irresistible' quality of the scientific talk at Caradale, Mitchison's house in

Scotland where 'the famous divide in the culture between science and the arts did not exist' and she almost certainly read Mitchison's science fiction, which addresses many of the themes subsequently explored by Lessing.[15] Mitchison's *Memoirs of a Spacewoman* (dedicated to the embryologist Anne McLaren) focuses on the influence of the maternal environment on the foetus, while *Solution Three* (dedicated to James Watson) explores reproductive cloning and the feasibility of a genetically planned society. Lessing's familiarity with such ideas meant that she was primed to respond to the sociobiology controversy when it hit the headlines in the 1970s. In common with most readers, Lessing is unlikely to have read the entirety of Wilson's 700-page tome *Sociobiology* but she may well may have read the controversial last chapter (on human evolution) and also Dawkins' shorter *The Selfish Gene*. Moreover, as Ullica Segerstråle has pointed out, the key sociobiological arguments were disseminated through extended critical debates in the media, for example in magazines such as the *New Scientist*, which ran a series of articles on the sociobiology controversy in 1976 and again in 1978, with contributors including E.O. Wilson and Stephen Jay Gould.[16] The same magazine, which Lessing cites as an important influence on her fiction, published a chapter of *The Selfish Gene* shortly after first publication, while in 1976 Dawkins and John Maynard Smith also appeared in a BBC *Horizon* programme entitled 'The Selfish Gene'.[17]

---

[15] Lessing recalls one incident in particular from her visits to Caradale, when she was asked to take a young, tongue-tied scientist for a walk: 'For about three hours we walked about over the hills and through the heather... At the end of it, wanting only to escape, I at last heard human speech. "The trouble is, you see, that there is only one other person in the world I can talk to". I reported this to Naomi, and we agreed that it was as dandified a remark as we could remember, even from a very young man. Quite soon he and Francis Crick would lay bare the structure of DNA.' See Doris Lessing, *Walking in the Shade: Volume Two of my Autobiography, 1949–1962* (London: Flamingo, 1998), p. 113. James Watson's memoir *The Double Helix* is dedicated to Mitchison.

[16] Ullica Segerstråle, 'An Eye on the Core: Dawkins and Sociobiology', in *Richard Dawkins: How a Scientist Changed the Way We Think*, ed. by Alan Grafen and Matt Ridley (Oxford: Oxford University Press, 2006), pp. 75–97 (pp. 78–9).

[17] See Doris Lessing, *Putting the Questions Differently: Interviews with Doris Lessing, 1964–1994*, ed. by Earl G. Ingersoll (London: Flamingo, 1996), pp. 42, 164. Discussing her space fiction in an interview with Thomas Frick, Lessing comments 'I know people say things like, "I regard you as rather a prophet." But there's nothing I've said that hasn't been, for example, in the *New Scientist* for the last twenty years' (p. 164).

As a style of thought, sociobiology's grand narrative is that of humanity as a species which is constrained by genes that prevent it from realizing the higher aims it can nonetheless envisage. This modern version of the Fall myth, in which original sin is transposed into genetic constraints, is the starting point for the fictional account of human evolution in Lessing's five Canopus novels. In the first, *Shikasta*, human history from palaolithic times to the late twentieth century is filtered through the perspective of the Canopeans, the most enlightened of three galactic empires who compete to control the earth (known as Rohanda/Shikasta) and its resources. The Canopeans see themselves as 'benevolent colonialists' yet their mode of governance involves both continuous surveillance and biopolitical control. They monitor the evolution of the planet and when humans emerge and promise to develop into a 'grade-A species', a more advanced species is introduced to tutor the humans and speed up their cultural evolution, while genetic improvement is brought about by manipulation of the gene pool. Shikastan children are only born after careful deliberation and 'from parents known to be the best', a strategy which echoes the 'conventional eugenics' envisaged by E.O. Wilson, and when a 'cosmic accident' compromises the success of this strategy, the Shikastans are given a direct 'boost' of superior Canopean genes. As the narrative point of view is that of the Canopeans, there is no explicit critique of these strategies: the critique only emerges in the third novel in the sequence, with the painful admission that corrupt populations have in the past been eliminated. However, Lessing draws attention to their unintended consequences. The Canopeans come to realize that their interventions are in part responsible for the ungovernable nature of the Shikastans and recognize too that although 'genetic prods' can have useful effects they can lead to genetic instability.[18] The concept of genetically-informed social planning is thus critiqued in its own terms, as it fails to deliver stability.

Lessing issues a further challenge to the idea of sociobiological planning by underscoring the parallel between the consequences of the Canopean interventions and the damage inflicted on the planet by Shikastans/humans in the present time of the narrative. In these

---

[18] Doris Lessing, *Shikasta*, p. 127.

sections of the novel, the narrative mode shifts from speculative fiction to contemporary realism as Lessing invokes the ecological damage inflicted by twentieth-century science and technology, as 'the great reservoir of nature' is poisoned by human stupidity (251). In this respect, she also introduces an alternative perspective on evolution, which is the view that human activity is introducing genetic mutations which may have damaging long-term evolutionary consequences. During the Cold War, fears about the genetic effects of atomic radiation were expressed by a number of prominent scientists, including Hermann Muller and Peter Alexander. However the case for more widespread damage (including genetic damage) was most eloquently put in Rachel Carson's *Silent Spring*, the book which has been credited with kickstarting the ecological movement. Lessing takes extracts from Carson's *The Edge of the Sea* as epigraphs for two of her earlier novels and *Silent Spring*'s account of ecological deterioration reverberates through *Shikasta*.[19] Arguing that 'genetic deterioration through man-made agents is the menace of our time', Carson emphasizes the mutagenic qualities of chemical agents as well as the better-known effects of radiation, noting that they 'disturb the cell's vital processes'.[20] She is outspoken about the politics of modern science, underscoring the link between biochemical research and agricultural profit, but it is her account of the ecological risks associated with scientific reductionism which intersects most closely with Lessing's interests. For Carson, the fabric of life is 'on the one hand delicate and destructible, on the other miraculously tough and resilient, and capable of striking back in unexpected ways' (257). In her view 'the extraordinary capacities of life' are threatened by a scientific attitude which she describes as 'primitive' in that it fails to take into account the interdependency of living systems and assumes instead that 'nature exists for the convenience of man'. Carson links such an assumption with 'the Neanderthal age of biology and philosophy', a comment that resonates with Lessing's decision to

---

[19] The extract that prefaces Part IV of *The Four-Gated City* emphasizes that the seemingly solid external landscape is composed of innumerable microorganisms which are always changing. The extract that prefaces *Briefing for a Decent into Hell* stresses the interconnectedness of living systems from the micro to the macro level.

[20] Rachel Carson, *Silent Spring* (London: Penguin, 2000), p. 189. Further page references will be given within the main text.

locate Canopean interventions in the era of prehistory, the implication being that they too belong to 'primitive' science (256–7).

Lessing broaches the question of scientific reductionism (to which Wilson and Dawkins subscribed) through her invocation of the Sufi perspective on evolution, which she first encountered when she read Idries Shah's *The Sufis*; as is well known, she subsequently became Shah's pupil and promoted his work in a number of prefaces and articles.[21] Sufism had an impact on the form as well as the content of her writing, as the Sufi teaching story impressed her as having the ability to influence 'the deepest and most hidden part of a human being'.[22] What attracted her was the hybridity of a genre in which (as in science fiction) the protagonists combine mimetic and representative traits, while the narrative blends elements of realism, myth, and folk tale. Such a mixing of narrative modes is evident in 'The Islands', the opening fable of Shah's *The Sufis*, which incorporates elements of sacred literature, utopian writing, and evolutionary theory. The narrative arc of this fable may have fed directly into the structure of the Canopus novels, as it features an 'ideal community' which is faced with an unspecified threat and relocated to an environment where it must adapt to coarser conditions.[23] After this 'Fall' the people are tutored by those few individuals who have been able to retain more 'advanced' qualities, and after many trials are ready to return to their former home. The story draws on the Fall myths of Christianity, Judaism, and Islam and reflects Shah's view that in striving for self-transcendence, human beings can further humanity's evolution. Such a progressive view of evolution is strikingly at odds with Darwin's cautious approach to teleology and is far from the pessimism or 'fatalism' that pervades sociobiology. Shah also departs from the determinism of sociobiology, placing far more emphasis on the interplay between the organism and the environment; particularly suggestive for Lessing was his claim that in an era when time and space 'are being transcended, humans are

---

[21] Many of these have been reprinted in Doris Lessing, *Time Bites* (London: Harper Perennial, 2005).
[22] Doris Lessing, Preface to the French Edition of *Seekers after Truth*, in *The Doris Lessing Reader* (New York: Alfred A Knopf, 1989), p. 634.
[23] Idries Shah, *The Sufis* (London: ISF Publishing, 2014), p. 1. Further page references will be given within the main text.

acquiring novel telepathic and prophetic powers' (67). In addition, and in line with the holistic philosophy of Sufism, he explicitly rejects scientific reductionism, arguing that complex entities cannot be explained in terms of the properties of their parts, and points to the limitations of a modern scientific methodology which takes no account of the subjectivity of the observer.

The Sufi perspective inflects *Shikasta* in that the Canopeans are modelled on Sufi teachers, who according to Shah are linked to their pupils by a 'magnetic force'. The Canopeans are similarly connected to the Shikastans by a flow of energies and like Sufi teachers, function as spiritual guides for promising individuals. At the same time, they act as eugenicists for the Shikastan population and play a major part in the genetic uplift with which the novel concludes. Anticipating a nuclear holocaust, they scour the planet looking for Shikastans with valuable genetic traits and send them away for safety so that when the mass of the population is destroyed, those who survive constitute an elite which is in the vanguard of human evolution. *Shikasta* thus recasts the biblical narrative of apocalypse and redemption in terms of both Sufism and sociobiology, perspectives which cohere only to the extent that both are concerned with fostering conscious evolution. They differ not only in their degree of optimism but also in their understanding of the relationship between the material and the immaterial. Sufi philosophy is holistic: it is predicated on the belief that humanity is perfectible but that, in Shah's words, 'the perfection comes about through attunement with the whole of existence. Physical and spiritual life meet' (29). Sociobiology is a strictly materialist science, grounded in the view that mental events are reducible to physical processes in the brain. As Lessing's novel engages with these competing perspectives, it develops a position which is critical of sociobiology's scientific materialism and which can be aligned not only with Sufi holism but also with Alfred North Whitehead's account of the relationship of the mind and nature. Although there is no direct evidence that Lessing had read Whitehead, his work is extremely apt for elucidating the critique of scientific materialism which she develops in *Shikasta*. For Whitehead, modern scientific thought is grounded in the 'bifurcation of nature', that is, the division of the world into two sets of things, one composed of the fundamental constituents of the universe, which are invisible but known to science, and the other being composed of what we are aware of in perception.[24] In this conceptual schema, the qualities attributed to an underlying physical reality are considered to be primary, while those that are referred to the mind are thought to be secondary, on the grounds that we cannot be certain that they exist independently of the mind. As Isabelle Stengers points out, neo-Darwinian sociobiological arguments are predicated on this distinction and on the hierarchy of value associated with it, as they attribute ontological reality to invisible genes and discount the ontological reality of human perceptions and purposes.[25]

Whitehead challenges the exclusion of human perception from the category of nature, arguing that the qualities attributed to external nature and those relative to the mind have an equal ontological status.[26] As he puts it, 'for us the red glow of the sunset should be as much a part of nature as are the molecules and electric waves by which men of science would explain the phenomenon'. In making this case, he aligns himself with what he calls 'our immediate instinctive attitude towards perceptual knowledge which is only abandoned under the influence of theory'.[27] This invocation of instinctive knowledge points to an affinity between Whitehead's philosophy and the genre of the realist novel, which starts from the premise that truth can be discovered through the senses, so it is unsurprising that it is in those sections of Lessing's text which deal most directly with Shikastan/human experience that a critical perspective on materialism emerges. Exploring the consciousness of the Shikastans as they reflect on the damage caused by their 'primitive' science, Lessing forges a perspective which coheres with Whitehead's challenge to the bifurcation of nature and with his process ontology. For Whitehead, processes are not things in the process of changing, rather, things are abstractions from the process of change. The most fundamental units of being are the 'events of experience'

---

[24] Whitehead develops this idea in *The Concept of Nature* (Cambridge: Cambridge University Press, 1964), pp. 29–31.

[25] Isabelle Stengers, *Thinking with Whitehead: A Free and Wild Creation of Concepts*, trans. by Michael Chase (Cambridge, MA: Harvard University Press, 2011), p. 177.

[26] In this respect, as Steven Shaviro has argued, Whitehead's thought anticipates the speculative realist critique of correlationism, that is, the belief that we can never grasp an object in itself in isolation from its relation to the subject. See Steven Shaviro, *The Universe of Things: On Speculative Realism* (Minneapolis: University of Minnesota Press, 2014), pp. 2–10.

[27] Whitehead, *The Concept of Nature*, p. 29.

which Whitehead also calls 'actual occasions' and what seem to be enduring objects are made up of strings of such occasions. Moreover, nothing is given in advance, for '*how* an actual entity *becomes* constitutes *what* that entity is'.[28] From this perspective, the dichotomy between mind and nature disappears, as both are abstractions from actual events. In the novel, Lessing dramatizes the bifurcation of nature in the consciousness of the Shikastans, who apprehend the beauty of a fallen autumn leaf ('There it lies in a palm, a brilliant gold, a curled, curved, sculptured thing') while at the same time they analyse its component parts in terms of 'the resources of chemical and microscopic cell life' (254). At this point however, the narrator intervenes to offer a perspective which resonates with Whitehead's process philosophy. S/he contends that neither the leaf nor the tree from which it falls is a static entity, rather, it is a complex of occasions, each one of which has an equal ontological value: the tree is 'not a tree, but a fighting seething mass of matter in the extremes of tension, growth, destruction, a myriad of species of smaller and smaller creatures feeding on each other, always—that is what this tree is in reality, and this man, this woman, crouched tense over the leaf, feels nature as a roaring creative fire in whose crucible species are born and die and are reborn in every breath' (255).

The second Canopus novel, *The Marriages Between Zones Three, Four and Five*, is also predominantly realist in mode.[29] Its central theme is the meaning of reproduction, considered in both biological and cultural terms. The female sex is represented by Al Ith, the queen of Zone 3, one of the five 'zones' which surround Shikasta. The zones represent stages of Shikastan/human development and can be considered analogous to the stages of enlightenment in Sufi thought, although the parallel is not exact; the novel also invokes the myth of Persephone, the goddess of fertility who spends six months of each year

---

[28] Alfred North Whitehead, *Process and Reality*, p. 23.

[29] In the Preface to *Shikasta*, Lessing comments that in comparison with the first Canopus novel, *Marriages* 'has turned out to be a fable, or myth. Also, oddly enough, to be more realistic' (Lessing, *Shikasta*, 'Preface', np). In an interview with Christopher Bigsby she also linked the novel with a Jungian technique she had used to deal with her own chaotic 'love life', creating an imaginary landscape in which she placed male and female figures in various relationships. See Doris Lessing, *Putting the Questions Differently*, pp. 81–2.

in the underworld.[30] In response to infertility in both their zones, Al Ith is required to undertake a sacrificial marriage to Ben Ata, the king of Zone 4. The marriage signals the need for greater genetic interplay between Zones 3 and 4 but Al Ith's (literal) descent into Zone 4 can also be read as a descent into the regressive sex/gender landscape depicted in sociobiological texts. The sociobiological argument for innate sex differences rested on one key observation, which was that across many species, males have millions of sperm while females have a limited number of eggs. In consequence, Wilson and others contended, the genetic interests of the male are best served by inseminating as many females as possible, whereas those of the female require the selection of a partner with 'good' genes who will share the cost of rearing the young. From these different needs flow different dispositions, or as Wilson puts it in *On Human Nature*:

> It pays males to be aggressive, hasty, fickle and undiscriminating. In theory it is more profitable for females to be coy, to hold back until they can identify males with the best genes. In species that rear young, it is also important for the females to select males who are more likely to stay with them after insemination. Human beings obey this biological principle faithfully. (125)

Philip Kitcher examines this argument in detail in his analysis of sociobiology and demonstrates that it depends on multiple assumptions which are not spelt out by Wilson, nor are their implications pursued. He concludes that the argument is 'gerrymandered' and suggests that Dawkins' discussion of 'the battle of the sexes' is open to similar objections.[31] Dawkins' account also turns on the asymmetrical relationship males and females have to reproduction and the differences in behaviour which are supposed to ensue. He is more cautious than Wilson in extrapolating from animal to humans but in his discussion of male and female reproductive strategies deploys strikingly anthropomorphic terms. For example, the alternatives open to females are presented as a choice between the pursuit of 'domestic bliss' or of the 'he-man', while 'feminine coyness' and 'prolonged

---

[30] In Sufi thought, there are usually agreed to be seven, not five, stages of movement towards the divine, so if Lessing were creating an exact parallel, there would be seven zones surrounding Shikasta.

[31] Kitcher, *Vaulting Ambition*, pp. 173-4.

courtship or engagement periods' are said to be common among animals (149, 157). The rhetoric conducts an argument by analogy, inviting the reader to conflate animal and human behaviour and to infer that in humans, characteristics such as 'feminine coyness' have been selected by evolution.

Marriages tests such claims through a thought experiment in which the representative of an egalitarian society, Al Ith, is plunged into a society organized around sociobiological axioms. As she is inducted into the regressive gender codes of Zone 4, her reflections on the process serve to highlight the complex ways in which the biological and the social are co-implicated in the formation of gender identity. In this respect Lessing opens up a perspective which accords with the biologist and gender theorist Anne Fausto-Sterling's analysis of the dynamic processes that shape sex/gender. For Fausto-Sterling, sex/ gender are not determined, as they are for sociobiologists, by genes that have been selected by evolution. Rather, they are forged in a developmental dynamic that is highly sensitive to context and caught up in a continuous relay between the individual and the environment. She points out that physiological sex is multi-layered and involves the acquisition of chromosomal, gonadal, hormonal, brain, and genital sex, each of which is open to environmental influence; conversely, gender identity is not just a social construction but also has a physiological dimension. Pursuing this point, taking as her example the stereotypical preference of little girls for the colour pink, Fausto-Sterling argues that although such gender preferences and gendered traits are context- and culture-specific, this does not mean that they are somehow disembodied. Rather, psychological reward systems mediated by the chemical dopamine may consolidate preferences and traits which are then inscribed at a neurobiological level.[32]

Al Ith's experiences can be aligned with such a perspective on the forging of sex/gender identities. After spending a considerable time in Zone 4, she morphs from being a person confident in her personal and sexual autonomy into someone who not only dresses in pink but spends days on her appearance, following the rigid protocols of Zone 4, in order

---

[32] See Anne Fausto-Sterling, *Sex/Gender: Biology in a Social World* (Abingdon: Routledge, 2012), chapter 9 for a discussion of the development of gendered traits.

to emphasize 'breasts, or legs, or arms'.[33] However, this is not just about the social performance of gender. Rather, Al Ith speculates that there is a neurophysiological circuit that links this behaviour with the sexual craving she has for Ben Ata, reasoning that 'If one devoted one's energies to self-display, to the exact disposition of parts of one's body... then presumably that was enough to call forth this raging desire? Cause and result. An energy spent thus will be answered thus' (222). Moreover, as she broods on the abject nature of her desire and her need for Ben Ata to 'extinguish her, knock her out, sink her deep' (222), she underscores the connection between gender inequality and a configuration of female desire in which submission is eroticized and hence embodied.[34] She also explores the way in which her experience of motherhood is changed as she is caught up in the affective web that underpins the 'backward and crude' reproductive practices of Zone 4. For example, in accordance with the strict separation of spheres in Zone 4, her child is born among a 'clutch of women' whose concern for him not only disconcerts Al Ith but is transmitted somatically, so that she too becomes anxious and uneasy. Moreover, under their influence she develops a passionate and possessive attachment to her son, something she has never experienced before although she has several children in Zone 3. When her son is born she is 'thrilled through and through with the wildest emotions of love and possession': again, Lessing draws attention to the link between gender inequality and a particular aspect of gender identity, in this case mother-love as a form of compensatory identification (202). Through Al ith, Lessing thus stages a sophisticated analysis of the construction of sex/gender identity, critiquing the 'just so' accounts offered by sociobiology but avoiding the purely constructionist approach which came to dominate the feminism of the 1980s. In this respect, her fiction points towards the limitations

---

[33] Doris Lessing, *The Marriages Between Zones Three, Four and Five* (St Albans: Granada Publishing Limited, 1981), p. 222. Further page references will be given within the main text.

[34] Lessing's analysis here is in line with Catharine Makinnon's argument that the features of women's status as second class, such as 'the restriction and constraint and contortion, the servility and the display, the self-mutilation and requisite presentation of self as a beautiful thing', are made into the content of sex for women. ('Sexuality, Pornography and Method: "Pleasure under Patriarchy"', *Ethics* 99 (Jan. 1989) 314–46 (318).

of the conceptual opposition between essence and construction, presciently anticipating the 'biological turn' in twenty-first-century feminist thought.[35]

Al Ith's negative view of Zone 4 derives from her experience of a more enlightened society, which if not quite a feminist utopia, as some critics have claimed, builds on a long tradition of feminist science fiction in its assessment of alternative approaches to sex and reproduction.[36] In Zone 3, sex is subtle and playful, with sexual relationships not being confined to an exclusive pair bond but ramifying out into a complex network of attachments. Reproduction, however, is the product of 'care, and thought, and long careful choices', and the development of the child is understood in terms very different from those of Zone 4 (72). Rather than parsing development in strictly genetic terms, Zone 3 also allows for the importance of epigenetic factors, that is, the environmental factors that can modify the expression of genes. In this respect, we can detect the influence of Mitchison's *Solution Three*, a novel which draws attention to the impact of the maternal body on the developing child. However, Lessing's text significantly expands our understanding of the range of factors that can affect foetal life, taking into account the affective environment that surrounds mother and child. The importance of this is formally recognized in Zone 3 as from conception a child is assigned to both 'Gene-Fathers' and 'Mind-Fathers', the latter being chosen for their capacity to imbue the foetus with their 'high and fine' qualities:

> When she had been pregnant—and after what care, and thought, and long careful choices—in the past, she had, as soon as she had been sure, chosen as beneficial influences for her child, several men who, knowing why they were chosen and for what purpose, co-operated with her in this act of blessing and gracing the foetus. These men had

---

[35] Examples of the rapprochement between feminism and biology include Elizabeth Grosz, *Becoming Undone: Darwinian Reflections on Life, Politics and Art* (Durham, NC: Duke University Press, 2011); Shannon Sullivan, *The Physiology of Sexist and Racist Oppression* (New York: Oxford University Press, 2015) and Elizabeth A. Wilson, *Gut Feminism* (Durham, NC: Duke University Press, 2015).

[36] See Marsha Rowe,'If you mate a swan and a gender, who will ride?' in *Notebooks, Memoirs, Archives: Reading and Rereading Doris Lessing*, ed. by Jenny Taylor (London: Routledge and Kegan Paul 1982, pp. 191–205 (p. 197).

> a special place in her heart and in the annals of her Zone. They were
> Fathers of the children just as much as the Gene-Fathers were. (72)

Lessing here conceptualizes development in ways that resonate strongly with C.H. Waddington's concept of the epigenetic landscape, itself a metaphor with the potential to cross disciplinary boundaries and generate unexpected insights.[37] Waddington's account of the epigenetic landscape is based on his work on *Drosophila* (the fruit fly) which demonstrated that phenotypes could change in response to an environmental stimulus and that such changes could be 'genetically assimilated' and therefore inherited. With the recent expansion of research in molecular epigenetics, the concept of the epigenetic landscape has been significantly expanded, as it has been shown that many aspects of the environment can leave 'epigenetic marks' and that in this sense, the environment can become embodied.[38] Lessing's novel pushes towards such an understanding as she points to affinities between the populations of the zones and their external environment. For example, Al Ith's father (who is also the official Chronicler of her story) comments on the isomorphic relationship between the people of Zone 3 and their environment, noting that 'with us the eye is enticed into continual movement, and then is drawn back always to the great snowy peaks that are shaped by the winds and the colours of our skies. And the air tingles in the blood, cold and sharp', while he suggests, conversely, that the inertia

[37] Waddington first used the metaphor of the epigenetic landscape in 1939 to capture the complex of developmental processes that connects the genotype to the phenotype. See C.H. Waddington, *An Introduction to Modern Genetics* (London: George Allen and Unwin, 1939) and *The Strategy of the Genes* (London: George Allen and Unwin, 1957). It is perhaps no coincidence that Waddington was heavily influenced by the philosophy of A.N. Whitehead, which impelled him to think about development in terms of dynamic interactions or as 'a Whiteheadian type of interacting network, rather than a straightforward linear sequence of cause and effect of the classical materialist kind'. See C.H. Waddington, 'Whitehead and Modern Science' in John B. and David R. Griffin Cobb, Jr, *Mind in Nature: the Interface of Science and Philosophy*, available at https://www.religion-online.org/book-chapter/chapter-5-whitehead-and-modern-science-by-c-h-waddington/ [accessed 30 August 2016].

[38] Epigenetics in its current sense refers to processes by which gene function is modified without changes to the DNA sequence. Some are heritable and they are stimulated by non-genetic factors in the cell, the body, and the external environment. In one sense the modern molecular definition is narrower than Waddington's, in another it is wider, as the factors potentially involved in epigenetic modifications extend from environmental chemicals to the psychosocial environment.

of Zone 4 might be linked to the dull uniformity of the landscape (35). From this perspective the marriage between Al Ith and Ben Ata can be seen to multiply both genetic and epigenetic possibilities, as Ben Ata suggests when he gently touches his unborn child and reflects on the 'potentialities . . . for the unknown and the unexpected' that have been unlocked by his marriage. The zones clearly represent ascending levels of evolutionary progress, as indicated by the metaphorical associations of their respective landscapes: Al Ith's journey is from the 'flat, dispiriting' Zone 4 through a return to the 'wild vigour' of Zone 3, then on to the airy and iridescent realm of Zone 2 (35, 77). However, this teleological view of evolution also accommodates a sense of the value of contingency and variety in the context of the complex interrelations between the organism and its environments.

The question of evolutionary progress—or its absence—is one to which Lessing returns in *The Sirian Experiments*, a novel which reverts to the macroscopic perspective of *Shikasta* to map the vicissitudes of Rohanda/Shikasta from the point of view of the Sirian empire. This text is more abstract than *Shikasta* or *Marriages* and is primarily concerned with a critique of Sirian expansionism which develops through a long-standing dialogue between a Sirian administrator, Ambien 11, and her Canopean mentor Klorathy. Through Ambien's first- person narrative, we learn that the Sirians have carried out 'bio sociological' and genetic experiments on a huge scale. They have transplanted indigenous peoples from numerous colonies to Rohanda, where they are adapted for work on other Sirian planets where hard labour is required. From one point of view, the novel can be read as a critique of scientific materialism and the Sirians' objectification of the species they manipulate. However, its critique of biopolitics has a further resonance which Lessing flags up in the afterword to *The Making of the Representative for Planet 8*. She points out that her account of Scott's Antarctic expeditions in this afterword is of obvious relevance to *Making*, which depicts a planet which is freezing to death, but equally she insists that *The Sirian Experiments* was written 'as a direct result of nearly fifty years of being fascinated by the two British expeditions to the Antarctic'.[39]

---

[39] Doris Lessing, *Canopus in Argos: Archives, The Making of the Representative for Planet 8* (London: Jonathan Cape, 1982), p. 123. Further page references will be given within the main text.

The connection may seem opaque but it lies in the imperial context of the expeditions, which took place at a time when European nations were competing for Antarctica as if on a stage illuminated by what Lessing calls 'that new toy, the popular newspapers'. Lessing is interested in the ethos of the expeditions and in the 'lofty exalted emotionalism' associated with its dedication to 'England, God, science' (134). The scientific strand of the second expedition was fuelled by this logic. It consisted of a search for King Penguin eggs which hatch in the depths of the Antarctic winter, the rationale for this being the belief that they could provide evidence for the evolutionary link between reptiles and birds, a belief which in turn rested on the assumption that the King Penguin was the most primitive bird in existence. The positioning of the bird as primitive and archaic coheres with the colonial construction of the Antarctic as a blank space ripe for appropriation, so that this quest for origins can be seen as the product of an intersection between colonial ideology and the turn of the century neo-Darwinian science associated with Wallace and Weismann.[40]

By extension, *The Sirian Experiments* can be read as an exploration of the convergence between more recent iterations of colonial ideology and aspects of sociobiological thought. The Sirians have been seen by many critics as representatives of Western colonialism in general but the novel also invokes the specific colonial history of Southern Rhodesia, which Lessing left in 1949 and to which she returned for the first time in 1982. In her account of the 'Loombi Experiment' for example, the Sirian narrator stresses the way in which the development of this 'low-grade' species is deliberately stalled by the Sirians in order that they will accept their role as an all-purpose labour force. This reads like an allusion to the denial of education and training to black Southern Rhodesians, in a system which, in the words of a 1975 UNESCO Report, 'trains Africans to provide efficient service at lower levels while ensuring for Europeans a superiority designed to confirm a

---

[40] For a perceptive analysis of the colonial dimensions of this expedition, see Elizabeth Leane, 'Eggs, Emperors and Empire: Apsley Cherry-Garrard's "Worst Journey" as Imperial Quest Romance', *Kunapipi* 31, 2009. As noted in the Introduction, the term neo-Darwinism was coined by Darwin's friend George Romanes to refer to the version of evolutionary theory developed by Alfred Russel Wallace and August Weissmann, which emphasized natural selection as the major driver of evolution and ruled out the Lamarckian idea of the inheritance of acquired characteristics.

racial mythology'.[41] Similarly, the space-lifting of the Loombis from uncongenial planet to uncongenial planet resembles the forced movement of the black populations of Southern Rhodesia to the unfamiliar and agriculturally poor land of the reserves. Moreover, the Sirians use biopolitical rhetoric to legitimate their abuse of their subject populations, constructing them as 'degenerate', a term which not only invokes the broader history of eugenics but which in Southern Rhodesia referred specifically to white settlers who had fallen away from white standards and adopted an 'African' way of life, one which could lead to miscegenation.[42]

The relationship between sociobiology and the concept of race is complicated. In his discussion of race in *On Human Nature*, Wilson goes out of his way to state that 'most scientists have long recognized that it is a futile exercise to try to define discrete human races'. However, this observation is followed almost immediately by a detailed account of a study which claimed to find significant, genetically determined differences in behaviour between newborns of Chinese-American and European ancestry. In terms of the textual space devoted to the arguments for or against innate racial differences, the argument in favour is given far more rhetorical weight (48). Wilson also argues both here and in *Sociobiology* that xenophobia is a genetically-mediated trait which has been inherited from our hunter-gatherer ancestors, while the sociobiologist and zoologist David Barash suggested in the 1970s that the principle of kin-selected altruism would favour genes which promote cooperative behaviour towards racial 'kin' and aggression towards those who look different from ourselves.[43] The effect of such arguments was to lend an aura of scientific respectability to the racism that persisted in the post-war period, despite such initiatives as the 1950 UNESCO Statement on Race, which had stated unequivocally that race had no validity as a scientific category.[44] Lessing encodes such

---

[41] *Racism and Apartheid: South Africa and Southern Rhodesia* (Paris: The Unesco Press, 1975), 40. Available at http://unesdoc.unesco.org/images/0001/000161/016163eo. pdf [accessed 28 May 2017)].

[42] See Dane Kennedy, *Islands of White: Settler Society and Culture in Kenya and Southern Rhodesia, 1890–1939* (Durham, NC: Duke University Press, 1987), p. 173.

[43] David Barash, *The Whisperings Within: Evolution and the Origins of Human Nature* (London: Penguin, 1979), p. 154.

[44] *UNESCO and Its Programme, Vol. 3, The Race Question* (Paris: UNESCO, 1950).

racism in her representation of the Sirians' othering of other species, as when they assume that the Loombis, who are 'simian' in appearance and therefore lower down the phylogenetic tree, are incapable of reflecting critically on their own subjection: they are astonished to discover that the Loombis are aware of their history of subjection and have constructed complex rituals to encode it. The racial logic which informs the Sirian experiments is further underscored when Ambien 11 realizes that that the biopolitical exploitation of the subjects in her experiments is indistinguishable from slavery.[45] Finally, Lessing makes the point that the enslavement of indigenous people frequently entails the destruction of non-modern knowledge which is both valid and irreplaceable, because it is possessed by those who have 'evolved from the earth and air and liquid of [the] planet'. Their medicine is 'based on plants and on psychological understanding' and they know 'how to live in and with the terrain in such a way that this [is] not damaged' (p. 260).

Ambien 11 comes to repudiate the Sirian experiments and campaigns to eliminate them, but what is equally important to the overarching argument of the Canopus novels is a related shift in her conception of scientific knowledge. Reflecting on her growing alienation from the Sirian worldview, she reflects that '*Facts*, the more experienced one became, were always to be understood, garnered, taken in, with that part of oneself most deeply involved with *processes*, with life as it worked its way out'(186–7). This signals an important transition in her thinking, from a conception of the world as being composed of things to one in which it is composed of processes. Through Ambien's critical reflections on the Sirian understanding of scientific progress, Lessing reiterates the view that objects of scientific enquiry should be conceived in terms of processes akin to Whitehead's 'actual occasions' or events. Such occasions or events are complex: as Steven Shaviro explains, each entity is different, and separate from all the others but they are also 'active and articulated processes—experiences and moments of feeling—rather than simple, self-identical substances'.[46] For Whitehead, every event

---

[45] Doris Lessing, *The Sirian Experiments: the Report by Ambien 11, of the Five* (London: Flamingo, 1994), p. 260. Further page references will be given within the main text.

[46] Steven Shaviro, *The Universe of Things: On Speculative Realism* (Minneapolis and London: University of Minnesota Press, 2014), p. 3.

contains some reference to every other event in the universe and its character is determined by the way in which it relates to everything else. Such an understanding of the dynamic nature of the universe and of the interrelatedness of entities across space and time is articulated by Ambien 11 when she comments that 'real scientists' (as opposed to the Sirians) are those who know how to match 'the ebbs and flows of the currents of life with invisible needs and imperatives' (291).

This perspective is further elaborated in the next novel in the sequence, *The Making of the Representative for Planet 8*, which maps 'a universe that is all gradations of matter, from gross to fine to finer, so that we end up with everything we are composed of in a lattice, a grid, a mesh'. In this short text Lessing returns to Canopus and an experiment in which the Canopeans create a new species by combining 'stock originating from several planets' (Lessing is never explicit about the nature of the Canopeans' genetic science but it seems to entail eugenic breeding strategies rather than direct manipulation of the genome). The species is healthy and well-adapted but the advent of an ice age brings the prospect of the death of the planet and the extinction of all species. Although the Representatives are not responsible for this catastrophe, their situation resembles that of twenty-first-century humans confronted with the evidence of anthropogenic climate change, and in mapping their response Lessing develops a perspective that intersects with current thinking about the Anthropocene. Specifically, it resonates with the post-anthropocentric perspectives developed by philosophers such as Rosi Braidotti, which deconstruct the notion of human nature as categorically distinct from the life of animals and non-human nature. Drawing on the language of evolutionary biology, Braidotti suggests that the Anthropocene demands a reconfiguration of human subjectivity based on the recognition that humans are 'environmentally based, that is to say embodied, embedded and in symbiosis with other species'.[47] Such a repositioning of the subject is at the heart of this novel, in which through the 'Representatives' Lessing articulates a view of the universe in terms of nested interrelationships stretching from atoms to stars, in a flattened ontology which makes no distinction in kind between human, animal, and geomorphic entities.

---

[47] Rosi Braidotti, *The Posthuman* (Cambridge: Polity Press, 2013), p. 67. Further references will be given within the main text.

Here human awareness is not privileged but is understood to be one among many forms of sentience, as stones are endowed with movement and intuition (they are described as 'a dance and a flow') and every cell and molecule of the body is understood to have 'an energetic and satisfactory life of its own' (65). In this respect Lessing is again close to Whitehead, whose panpsychism has been the most controversial aspect of his philosophy. For Whitehead, each entity from a rock to a neutrino is an autonomous centre of value: in his view—as in Lessing's—it is crucial to emphasize that each entity has value for itself, regardless of the human point of view.[48] In this context of a decentring of the human, Lessing renders the death of the individual and the species in terms of transitions from one level of organization to another, from an existence as 'webs of matter or substance or something tangible, though sliding and intermingling and always becoming smaller and smaller' (118). In representing death as a merging with the cosmos, Lessing's fiction coheres with Braidotti's account of death as 'becoming imperceptible . . . the point of evacuation or evanescence of the bounded selves' (137). Braidotti suggests that in understanding both life and death as expressions of *zoe*, we can decentre the ego and collectively invest in *zoe* as a force which extends beyond the self. Death becomes an affirmation of our relationship with the inhuman, a perspective which is dramatized in the Representatives' farewell to their dead planet: 'as we swept on there, ghosts among the ghostly worlds, we felt beside us, and in us, and with us, the frozen and dead populations that lay under the snows . . . [they] were held there for as long as the ice stayed, before it changed, as everything must, to something else—a swirl of gases perhaps, or seas of leaping soil' (120–1).

Of the five Canopean novels, *Making* and *Marriages* have had the most positive critical reception and have also been turned into operas in a collaboration between Lessing and the composer Philip Glass.[49] The concluding novel, *Documents Relating to the Sentimental Agents in*

---

[48] For Whitehead, all entities are active, intentional, and vital, and each has a 'sense of existence for its own sake, of existence which is its own justification'. Alfred North Whitehead, *Modes of Thought* (1938), quoted in Shaviro, *Universe of Things*, p. 89. The further implication is that thought, value, and experience are not exclusively human properties.

[49] The opera of *The Making of the Representative for Planet 8* was first performed in 1988, and that of *The Marriages Between Zones Three, Four and Five* in 1997.

the Volyen Empire, is the one that has been least well-received. Lessing has acknowledged that she 'got lost' in writing it, caught up in the pleasure of satirizing the kind of political rhetoric she was familiar with from her time in the Communist Party.[50] For the purposes of this discussion, the most significant feature of the novel is its exploration of the tensions between sociobiological and Marxist/constructionist perspectives on human history. In this respect, the novel can be read as a commentary on the debate which broke out after the publication of *Sociobiology* when Wilson was attacked by a group known as the 'Sociobiology Study Group of Science for the People'. This included the geneticist Richard Lewontin and the paleontologist Stephen Jay Gould, both of whom were on the left politically and had Marxist sympathies. The group focused on the lack of evidence for the claims Wilson made for human behaviour and argued that his ostensibly objective scientific project concealed an intention to defend 'the status quo as an inevitable consequence of "human nature"'.[51] Similar points were made in Steven Rose, Richard Lewontin, and Leon J. Kamin's *Not in our Genes*, which attacked what it saw as a 'rising tide of biological determinist writing' which explained inequalities through a reductionist theory of human nature.[52] As Kitcher has demonstrated, the critique of sociobiology was entirely valid from a scientific point of view because in contrast with Wilson's entomological research, his claims about human nature were poorly evidenced and his methodology was problematic. Yet what caught the attention of the public was not the detail of the scientific debate but the conflict between biological and constructionist models of human behaviour. This was a period when social constructionism dominated in the social sciences and biological explanations of class, gender, and race were viewed with suspicion, but it was also a time when new theories about the evolution of human behaviour were gaining traction in academia and in the popular imagination.[53]

[50] Lessing, *Putting the Questions Differently*, p. 170.

[51] Allen, Elizabeth, et al., 'Against "Sociobiology"', *The New York Review of Books*, 13 November 1975 http://www.nybooks.com/articles/1975/11/13/against-sociobiology/ [accessed 28 May 2017].

[52] Steven Rose, R.C. Lewontin, and Leon J. Kamin, *Not in our Genes: Biology, Ideology and Human Nature* (London: Penguin, 1984), p. ix.

[53] Popular books by Robert Ardrey and Desmond Morris were important in brining ideas about the biological origins of human behaviour to a wider public. Ardrey's *The*

In *Sentimental Agents*, Lessing creates a critical distance from both perspectives by drawing attention to the epideictic (declamatory) rhetoric used by their exponents. Marxist rhetoric is the object of direct attack in the novel, as the standard counters of Marxist analysis are ridiculed by a Canopean agent and held up as examples of 'rhetorical disease': key terms include the 'winds of history' and the 'logic of history'.[54] However, Lessing also draws attention to the use of epideictic rhetorical techniques in sociobiology, as another Canopean offers a sociobiological perspective on human development and uses the rhetorical technique of epistrope to persuade his audience that 'Long ago, in your animal and semi-animal past, past, you were groups and bands and packs, and on these genetic inclinations tyrants have built, to keep you in groups and bands and packs; but now you can free yourselves, because you understand yourselves...' (131). Wilson and Dawkins employ similar if more understated rhetorical techniques, being particularly fond of antithesis. In drawing attention to the epideictic dimension of sociobiological writing, Lessing underscores its polemical quality and its imbrication with wider political debates which were to come to greater prominence in the 1990s with the so-called science wars, in which there was a standoff between scientific and humanistic views of human nature from which the humanities have arguably not yet recovered.[55]

As Lessing probes the assumptions and values which underpin the sociobiological enterprise, she also reassesses the humanist values which were at the core of her early writing. Her commitment to these values is articulated in the well-known essay 'The Small Personal Voice', in which she aligns herself with the tradition of the realist novel which is 'the highest form of prose-writing' because of its

---

Territorial Imperative: A Personal Inquiry into the Animal Origins of Property and Nations was published in 1966, while Morris's *The Naked Ape: A Zoologist's Study of the Human Animal* was published in 1967.

[54] Doris Lessing, *Documents Relating to The Sentimental Agents in the Volyen Empire* (London: Flamingo, 1994), p. 35. Further page references will be given within the main text.

[55] See Chapter 2 for further discussion of this point. See also John Guillory, 'The Sokal Affair and the History of Criticism', *Critical Inquiry* 28 (Winter 2002), 470–508 for an assessment of the response of the humanities to the 'science wars'.

commitment to 'warmth and humanity and love of people': any writer with a sense of responsibility, she writes, 'must become a humanist'.[56] The argument of the essay also overlaps with Lukacs' Marxist humanism and his understanding of literary character as a concentrated expression of the forces of history. The broader Marxist analysis of history was formative for Lessing, whose thinking is constitutively grand in scale, hence perhaps her subsequent turn to sociobiology, which offers an account of human history which is even more capacious than that of Marxism in that it stretches back to 'deep' or unrecorded history. However, as Lessing's fiction probes sociobiological propositions and tests them against the contingencies of experience, it opens up perspectives which run counter to sociobiological reasoning and the scientific humanism which underpins it. This was the term coined by E.O. Wilson to denote an evolutionarily informed humanism which positions humans as rational subjects struggling with 'predispositions' which date to an earlier stage of evolution, a perspective which is shared by Richard Dawkins. The humanism of both thinkers is thus grounded in a species hierarchy which assigns the capacity for rational thought only to humans and it is one in which the natural world (both organic and inorganic) is construed as lacking in agency, a passive backdrop for the autonomous human subject. In challenging these assumptions, Lessing's fiction anticipates a number of aspects of post-humanist thought and in particular, as we have seen, it can be aligned with the post-anthropocentric aim of decentring the human and recalibrating the relationship between humans and other species. To this extent, her fiction foreshadows the transition which structures the contemporary post-anthropocentric moment, as traditional concepts of human agency and responsibility are being called in question by thinkers like Dipesh Chakrabarty, who argues for a conceptualization of the human at the level of the species, and Timothy Morton, who has called for a reconfiguration of the concept of 'nature' and for new ways of imagining ecological co-existence.[57]

---

[56] Doris Lessing, *A Small Personal Voice: Essays, Reviews, Interviews*, ed. by Paul Schlueter (New York; Alfred A Knopf, 1974), p. 6.

[57] Dipesh Chakravarty makes this case in 'The Climate of History: Four Theses'. For Timothy Morton's arguments see *Ecology without Nature* (Cambridge, MA: Harvard University Press, 2009) and *Dark Ecology: For a Logic of Future Coexistence* (The Wellek Library Lectures), New York: Columbia University Press, 2016).

In opening up these questions, Lessing's fiction takes a rather different direction from that of E.O. Wilson, whose engagement with ecological issues dates to the publication of *Biophilia* in 1984, just a year after Lessing's *Sentimental Agents*. Wilson's concept of biophilia is premised on the sociobiological assumption that 'the whole process of our life is directed towards preserving our species and personal genes'.[58] He goes on to argue that biophilia, which he defines as an attraction towards living beings, is innate and genetically hardwired in humans, having been selected because of its adaptive advantage. As noted above, Wilson's perspective is predicated on a species hierarchy which positions humans at the apex on the grounds of their unique cognitive ability. This is a position he maintains in the 1986 essay collection he co-edited, *The Biophilia Hypothesis*, but in another contribution to this collection, Dorion Sagan and Lynn Margulis offer a perspective which can be aligned more readily with contemporary post-anthropocentrism. They argue that the range of responses between organisms (including those between humans and non-humans) is more appropriately described in terms of 'prototaxis', which is often a precursor of symbiosis, rather than biophilia.[59] They offer a strong critique of the human exceptionalism which informs Wilson's argument and suggest that the idea that humans can by themselves save the planet is itself a form of species arrogance. For Sagan and Margolis, the prospect of human extinction must be placed in the context of a global ecosystem which 'is far bigger and more metastable than any single life-form' (363). From such a perspective, humanity is dispensable, a perspective Lessing endorses in *Shikasta* when at the point when the Shikastans are forced to confront the possibility of their own extinction, the narrator writes that they have come to appreciate that 'the universe is a roaring engine of creativity, and they are only temporary manifestations of it' (256).

[58] Edward O. Wilson, *Biophilia* (Cambridge, MA: Harvard University Press, 1984), p. 120.

[59] Dorion Sagan and Lynn Margulis, 'God, Gaia, and Biophilia', in *The Biophilia Hypothesis*, ed. by Stephen R. Kellert and Edward O. Wilson (Washington, DC: Island Press, 1993), pp. 345–64 (p. 347). Margulis was the first to suggest that eukaryotic cells were formed as the result of a merger between different forms of bacteria. With James Lovelock, she was the co-developer of the Gaia Hypothesis. Further page references will be given within the main text.

2

A.S. Byatt's Biological Reason

Discussions of A.S. Byatt's interest in science usually focus on the last two novels of the sequence known as the *Quartet*, in line with her description of this as having changed from 'a backward glance at the power of Shakespeare's and Milton's English and England, to a form excited by the mystery of scientific discovery'.[1] However, Byatt's intellectual interests have always been wide-ranging and from the outset the *Quartet* displays an interest in the foundations of thinking across disciplines, in numbers, geometry, and grammar. The ambition of the sequence is to invoke the intellectual climate of the recent past and to map the history which lies behind it, creating something like a fictional counterpart of the archaeology of knowledge which Foucault develops in *The Order of Things*. This is a text to which Byatt frequently refers in her essays and Foucault's history of the successive *epistemes* which have shaped Western thought is a recurring point of reference in her fiction. Accordingly, this chapter reads *The Virgin and the Garden* and *Still Life* in the light of Foucault's analysis of biology as a component of the modern *episteme*, before tracing Byatt's subtle and intricate engagement with neo-Darwinian theory in *Babel Tower* and *The Whistling Woman*. In these later novels Byatt presses further on the philosophical issues raised by Foucault and examines the complicated

---

[1] A.S. Byatt, 'Fiction informed by science', *Nature* 435, 12 March 2005, 294–7. Further page references will be given within the main text.

Genetics and the Literary Imagination. Clare Hanson, Oxford University Press (2020). © Clare Hanson.
DOI: 10.1093/oso/9780198813286.001.0001

entanglement of biological and social factors within evolutionary theory. The chapter concludes with an analysis of *The Biographer's Tale*, a novel which explores the history of neo-Darwinism and considers the relationship between the life sciences and life writing. Byatt's approach is distinctive in that she engages directly with the scientific archive, often referencing scientific papers and edited transcripts of scientific debates. She is interested in the philosophical implications of neo-Darwinian theory and in its impact on our understanding of human relations: in exploring these issues her fiction provides a modern version of George Eliot's engagement with evolutionary theory.

The Virgin and the Garden is set in the 1950s but the action is viewed through the double lens of a frame narrative from the 1960s and narrative interpolations from the time of writing, a technique of temporal layering which Byatt uses throughout the *Quartet* to suggest the contingency of her protagonists' beliefs.[2] The novel evokes a literary culture beset by nostalgia for a 'Golden Age' of Elizabethan literature in which words and things were indissolubly linked. T.S. Eliot was the most influential exponent of this view through his theory of the dissociation of sensibility in the seventeenth century, but as Byatt points out, there is a suggestive analogy between his argument and Foucault's account of the Renaissance *episteme* in which linguistic signs were continuous with the language of nature.[3] Both writers suggest that language has subsequently become divorced from representation so that, as Foucault puts it, what language speaks is 'the word itself—not the meaning of the word, but its enigmatic and precarious being'.[4] Byatt's main protagonist, the literary Frederica Potter, is preoccupied

---

[2] For a discussion of this technique see Alistair Brown, 'Uniting the Two Cultures of Body and Mind in A.S. Byatt's *The Whistling Woman*', *Literature and Science* 1 (2007), 55–72. Brown argues that Byatt's undermining by anachronism 'ensures that science is contested in terms of other branches within science, and as part of a historically adaptable series of paradigm perspectives' (67).

[3] Foucault uses the term *episteme* to refer to the unconscious ideas of order which determine what is accepted as knowledge in particular periods. He argues that the Renaissance *episteme* was based on similitude and resemblance whereas the Classical *episteme* was grounded in relations of identity and difference. In the modern *episteme*, humanity became the main subject of the empirical sciences, which focused on invisible inner structures.

[4] Michel Foucault, *Les Mots et les Choses* (1966, first English translation Tavistock Publications, 1970), *The Order of Things* (Abingdon: Routledge, 2002), p. 333. Further references will be given within the main text.

with the relationship between words and things but the novel is equally interested in the relationship between abstract symbols and empirical reality in the sciences, a theme which is explored through the experiences of Frederica's brother Marcus. As a child, Marcus can solve mathematical problems by imagining a garden filled with forms: 'I used to see—to imagine—a place. A kind of garden. And the forms, the mathematical *forms*, were about in the landscape and you would let the problem loose in the landscape [...] then I saw the answer.'[5] These visions entail a sense of continuity with the external world which echo Foucault's description of the Renaissance *episteme*: as the narrator explains, when Marcus sees the garden, '*In* this garden Marcus was not exactly: rather, he was coextensive with it, his mind its true survey' (337).[6] He loses this ability after being questioned about it but continues to be haunted by the possibility that numbers inhere in things. By interweaving Marcus's story with that of his sister, Byatt draws attention to the way in which our understanding of the relationship between representation and reality fluctuates over time, both for the individual and in wider culture.

Against this backdrop, biology erupts into the novel in the shape of Lucas Simmonds, who teaches the subject at Marcus's school. Simmonds has concocted an eccentric theory of evolution which borrows from the work of V.I. Vernadsky, the Russian mineralogist who was one of the first to develop the concept of the biosphere. From Vernadsky, Simmonds takes the idea of human cognition as a power which will eventually transform the biosphere into the noussphere, an evolutionary stage which Vernadsky conceives as being dominated by scientific reason. However, Simmonds replaces Vernadsky's rationalism with a mysticism which owes a good deal to Teilhard de Chardin and claims that the transfiguration of Man into pure Mind signifies 'a working out of a higher purpose' and the 'realisation of a Plan' (147). For Simmonds, Marcus's mathematical gifts and synaesthesia-induced

---

[5] A.S. Byatt, *The Virgin in the Garden* (1978) (London: Penguin, 1981), p. 63. Further page references will be given within the main text.

[6] This sentence contains a buried allusion to Andrew Marvell's 'The Mower's Song', which begins 'My mind was once the true survey / Of all these meadows fresh and gay'. This takes us back to Eliot's concept of the dissociation of sensibility. Arguably, such a dissociation is enacted in Marvell's poem, which suggests that the pastoral scene it depicts no longer coheres with the reflective consciousness of the mower.

hallucinations suggest that he is an agent of the Plan and accordingly he recruits him for experiments in thought transference which bring both to the point of collapse. Simmonds's preoccupation with obscure energies, with change and decay, resonates with Foucault's account of the emergence of biology as a form of thought. For Foucault, when biology replaced natural history it inaugurated an 'untamed ontology' in which the 'flat' structure of Linnaean taxonomy was replaced by a focus on invisible inner forces. The spatial continuity of the Classical *episteme* gave way to a historical continuity predicated on distant evolutionary origins, while August Weissman's influential account of inheritance represented the individual organism as a temporary vehicle for the immortal germ plasm.[7] In this context, Foucault argues, life took on a radical value, becoming the nucleus both of being and non-being: as he writes, 'there is being only because there is life, and in that fundamental movement that dooms them to death, the scattered beings, stable for an instant, are formed, halt, hold life immobile [. . .] but are then in turn destroyed by that inexhaustible force' (303). From this point of view, life is implicated with death because it always outlives the individual organism; death is inseparable from such a genealogical conceptualization of life.

This genealogical perspective is explored in more depth in *Still Life*, a novel which, as Byatt notes in her *Nature* article, is directly concerned with 'biological life—sex, birth and death' (294). The novel approaches these questions by invoking Freud's 'Beyond the Pleasure Principle', a text which is framed by the biblical myth of the Fall, on the one hand, and nineteenth-century evolutionary theory on the other. Weissman reappears in this context, as Freud is struck by the analogy between Weissman's distinction between the immortal germ and the disposable soma and his own hypothesis of a struggle between the 'life instincts' and the death drive.[8] However, having registered Weissman's

[7] See August Weismann, *Essays upon Heredity* (Oxford: Oxford University Press, 1889), esp.org, http://www.esp.org/books/weismann/essays/facsimile/contents/weismann-essays-1-a-fm.pdf [accessed 27 November 2018].

[8] Byatt discusses Freud's interest in Weissman in her essay 'Faith in Science', where she suggests that he was fascinated by Weissman's formulation that 'while the body was mortal, the germ cells were potentially immortal'. See A.S. Byatt, 'Faith in Science', *Prospect Magazine* 20 November 2000, https://www.prospectmagazine.co.uk/magazine/faithinscience [accessed 26 November 2018].

significance as a biological theorist the novel goes on to explore the re-inscription of his theory of 'programmed death' in neo-Darwinian thought. Programmed death became a prominent issue in the 1950s and 1960s as biologists began to explore the relationship between reproduction and senescence. The distinguished immunologist Peter Medawar argued that ageing was caused by an accumulation of genetic mutations, while George C. Williams developed a theory of 'antagon-istic pleiotropy' which turned on the idea of genes which simultan-eously coded for reproduction in early life and ageing in later life: natural selection, he argued, would favour such genes.[9] Weissman's distinction was thus re-inscribed at the genetic level, giving additional force to the distinction between the immortal germ and the mortal body, as evidenced by Richard Dawkins's account of the gene as leaping 'from body to body down the generations, manipulating body after body in its own way and for its own ends, abandoning a succession of mortal bodies before they sink in senility and death'.[10] In *Still Life* this bleak view of human existence is strongly associated with Frederica's sister Stephanie. For example, when she is watching a nativity play the narrator comments that on such occasions parents are 'moved by some threat in the law of flesh and blood itself. These small creatures are the future, they are only acting out what they will be. Not only childhood vanishes: men and women, having handed on their genes, are super-fluous.'[11] The nativity is recast in neo-Darwinian terms, and in place of Christianity's promise of individual salvation the children are seen as doomed to repeat the cycle of reproduction and death, their genetic immortality being purely virtual because endlessly deferred. Stephanie's shocking death at the end of the novel acts as retrospective confirm-ation of the view that when 'the immortal life of the genotype is

[9] See George C. Williams, 'Pleiotropy, Natural Selection and the Evolution of Senescence', *Evolution* 11 (December 1957), 398–411.

[10] Richard Dawkins, *The Selfish Gene* (1976) (Oxford: Oxford University Press, 1989). As Dawkins explains, in the first edition of the book he failed to acknowledge Williams's work, emphasizing instead Medawar's contribution to the science of ageing (274). As he also notes, biologists in America were far more likely to encounter these ideas through Williams's work—an example of the distinctiveness of scientific culture in the UK and in the USA.

[11] A.S. Byatt, *Still Life* (1985) (London: Vintage, 1995), p. 50.

transmitted [...] so the phenotype, the individual body, becomes redundant' (286).[12]

The opposition between innatism and constructionism was fundamental to both the sociobiology controversy and subsequent debates prompted by evolutionary psychology. However, Byatt approaches this issue from an oblique angle, through Massimo Piatelli-Palmarini's *Language and Learning*, an edited version of the 1975 Royaumont symposium which featured Jean Piaget and Noam Chomsky. This book meant 'a great deal' to Byatt because it addressed the question of the acquisition of language, and its arguments feed directly into her exploration of the development of Stephanie's young son, Will. For Piatelli-Palmarini, the innatism/constructivism debate is one in which 'two opposing ontological commitments [...] compete to provide an understanding of how life is possible'.[13] One is what he calls the crystal model, which takes its name from Edwin Schrodinger's prediction that the gene would turn out to be an aperiodic crystal. This model assumes that biological organization can be explained in terms of intrinsic and invariant molecular structures: order at the macroscopic level is generated by order at the microscopic level and the environment possesses no features which can be assimilated by the organism. As Piatelli-Palmarini puts it, in this view 'all laws of order, whether they are biological, cognitive, or linguistic, come from inside, and order is *imposed* upon the perceptual world, not *derived* from it' (1). This model coheres with Chomsky's theory of a universal grammar which pre-exists any interaction with the environment. In contrast, the 'flame' model assumes that living systems are constituted according to the principle of self-organization, in which order is formed from disorder, as in a flame which retains its integrity despite internal and external perturbations. In this model, the organism can reorganize itself in response to environmental cues, a view which accords with Piaget's emphasis on self-regulation. The crystal and flame opposition is initially presented in agnostic terms as Byatt's narrator comments that

---

[12] This point is given additional force as Stephanie's last thoughts are of her children, followed by 'the word altruism, and surprise at it' (p. 403). In this context, the word altruism hovers equivocally between its original and neo-Darwinian senses.

[13] Massimo Piattelli-Palmarini, ed., *Language and Learning: The Debate between Jean Piaget and Noam Chomsky* (Cambridge, MA: Harvard University Press, 1980), p. 6. Further references will be given within the main text.

'human cognition has been called "order from noise": or it may be, contrariwise, the patterning of the world with a constructed map, crystallised in the genes' (287). As the account of Will's development unfolds however, the narrative inclines towards the crystal position, which accords with Chomsky's Cartesian view that we acquire knowledge through the imposition of pre-existing ideas on the world. The example Descartes gives is that of a triangle which pre-exists any encounter with an instance of it and in line with this idea, Will interprets his early experiences in terms of geometric form: 'when he began to draw he would draw five flat loops round an approximate circle until he learned the pleasures of the compass-constructed over-lapping arcs which make up, in their way, a Cartesian flower—a Platonic Flower?' In moving from the idea of a Cartesian flower to a Platonic flower, the narrator makes a suggestive connection between hard-wired biological structures and the Platonic ideals which pre-occupy Frederica in both *The Virgin in the Garden* and *Still Life*.

Babel Tower takes up these and related issues through the deliberations of a government committee on the teaching of language in schools.[14] The committee members fall into two camps.[15] On the one hand are those who believe that children need to be liberated from restrictive grammatical rules, their main representative being the Liverpool performance poet Mickey Impey. This group takes the anti-foundationalist, constructionist position articulated by Piaget in *Language and Learning*, overlaid with a distinctly 1960s optimism about the possibility of remaking language in the image of freedom. On the other side is a group led by Gerard Wijnnobel, a mathematician, grammarian, and vice-chancellor of the University of North Yorkshire who believes that children should be taught the forms of language so that they are able to describe the structure of their thought. In support

[14] In the twelve-year interlude between *Still Life* and *Babel Tower* Byatt wrote the Booker prize-winning *Possession*, which also engages with theories of language, specifically those associated with poststructuralism. Through the character of Roland Michell, the novel offers a critique of the view which dominated literary theory in the 1980s, that language 'could never speak what was there, that it only spoke itself'. A.S. Byatt, *Possession* (1990) (London: Vintage, 1991), p. 473.

[15] For this fictional committee Byatt drew on her experience of serving on the Kingman Committee of Inquiry into the Teaching of English Language and on her own memories of the hostility to science which pervaded the counter-culture of the 1960s. See Byatt, 'Fiction informed by Science', 296.

of his view, Wijnnobel invokes Chomsky's theories of generative and transformational grammar and Nietzsche's claim that ideas are formed by grammatical functions with a biological origin.[16] For Wijnnobel, the grammatical forms and structures we use are 'part of the structure of our brains informed by our genes' and both the 'extraordinary subtlety' and the limitations of human intelligence are a function of this innate structure.[17] Moreover, 'worrying at insoluble "problems"' may be a function of the limits of human intelligence, the implication being that the innatism/constructionism debate might be an instance of this. In making this point Wijnnobel draws attention to the crucial question Foucault raises in *The Order of Things* of the 'analytic of finitude' and the problem this presents for modern philosophical thought. Foucault starts from the proposition that the modern empirical sciences describe man as a finite being who is limited by the biological and linguistic forces which as Foucault puts it 'overhang him [...] and traverse him as though he were merely an object of nature' (341-2). However, if man is a historically limited empirical being, how can he simultaneously be the source of the representations by means of which he makes sense of the empirical world? To put it another way, how can 'I' be both an empirical object of representation and the transcendental source of representations? In Foucault's view, this is an aporia which cannot be resolved and which will ultimately bring about the collapse of the modern *episteme*. He contends that positivist science sidesteps the issue by arguing that 'knowledge has anatomo-physiological conditions, that it is formed gradually within the structures of the body', thereby reducing the transcendental to the empirical (347). Wijnnobel exemplifies this position, as he believes that thought can ultimately be explained in terms of the genes which dictate 'the folds of the cortex, the dendrites and synapses and axons of the neurones in the brain' (192).

Such positivism also underpins the work of Steve Jones, the geneticist whom Byatt acknowledges as an important source for her fiction. Jones had worked on genetics in snails for decades at the time when Byatt met him and had also just given the 1992 Reith Lectures, subsequently published as *The Language of the Genes*, one of many books which

---

[16] Byatt here draws on Nietzsche's thinking in *The Will to Power* while the epigraph to *Babel Tower* comes from *Twilight of the Gods/The Anti-Christ*.

[17] A.S. Byatt, *Babel Tower* (1996) (London; Vintage, 2003), p. 186.

mobilized the trope of genetics as 'the language of life'. As Lily Kay has shown, this metaphor was not simply used to appeal to a wider audience, as linguistic representations were to some extent integrated into the interpretive and experimental framework of geneticists.[18] Nonetheless, as metaphor, it has the effect of anthropomorphizing DNA, naturalizing it by assimilating it to culture. Jones is sensitive to the importance of culture and is at pains to emphasize the impact of the environment on genes. However, he sees the relation between them in terms of an interaction between discrete and bounded entities rather than between components of a single dynamic system, a perspective which excludes the possibility of self-organization and discounts the significance of non-genetic forms of inheritance. Byatt borrows Jones's research and gives it to Luk Lysgaard-Peacock, a fictional representative of neo-Darwinism whose views closely resemble Jones's, being positioned at a mid-point on the continuum between innatism and constructivism. However, if Luk represents the genetic middle ground, his student Jacqueline Winwar diverges from genetic orthodoxy when she develops an interest in the engram, a term introduced in 1904 by Richard Semon to describe the neural substrate which stores and retrieves memories. Semon defines the engram in terms of a physical change in brain state in response to an event or experience, a change which could subsequently be reinforced by similar events. His work was largely ignored by his contemporaries, probably because he argued that learning acquired by a parent could be passed to future generations, thus endorsing 'soft' or Lamarckian inheritance.[19] However, the 1960s saw renewed interest in the engram and in so-called 'molecules of memory', stimulated by the work of James V. McConnell, who thought that memory might be encoded in RNA. McConnell's somewhat bizarre experiments involved training flatworms to avoid bright light which they associated with electric shocks. When they were subsequently bisected, each half retained an aversion to bright light; moreover, when the bisected worms were chopped up and fed to untrained

---

[18] Lily E. Kay, *Who Wrote the Book of Life? A History of the Genetic Code* (Stanford: Stanford University Press, 2000), pp. 17–18.

[19] For a detailed historical account of the engram and its reception see Sheena A. Josselyn, Stefan Kohler, and Paul W. Frankland, 'Heroes of the Engram', *Journal of Neuroscience* 37, 4647–57 (3 May 2017) doi:10.1523/jneurosci.0056–17.2017.

worms, the latter also avoided bright light. From this McConnell concluded that memory was retained and transferred at the molecular level and that it could even be transferred between organisms. Like Semon he was a Lamarckian, swimming against the tide of molecular genetics, and the validity of his experiments was widely questioned. Lyon Bowman, the physiologist who describes his research in the novel, reflects this scepticism when he jokes 'what fear of the butcher, what desire for grass pastures, should I not have absorbed from my steak and kidney?' (250). Nonetheless the idea that learned information could be transmitted via DNA or RNA becomes the focus of Jacqueline's work, which is based on the supposition that 'learned information, as well as genetic coded information, might be retained in and transmitted by very large molecules, such as the DNA and the RNA' (252). This suggests the further possibility that DNA could be modified in response to environmental cues, which if true would contravene the 'Central Dogma' of molecular biology, that DNA codes for RNA codes for protein in a unidirectional way. Byatt thus invokes an alternative to the neo-Darwinism espoused by most of her scientist-protagonists and gestures to the possibility that DNA might itself be modifiable or plastic, although this is not a theme which is pursued in detail; rather, the novel privileges the neo-Darwinian view of genes as fixed and determining.[20]

This does not prevent Byatt from being critical of the extreme version of genetic determinism which was associated with the intersection of cybernetics and molecular biology. To develop this critique, Byatt draws on Jean Dupuy's account of the Macy conferences on cybernetics in his *The Mechanization of Mind* (1994).[21] These conferences were attended by mathematicians, biologists, embryologists, and psychologists, all with an interest in complex regulatory systems.

---

[20] As Josselyn et al. note, recent research has confirmed that changes in DNA methylation (i.e. epigenetic changes) are implicated in learning and memory, taking the story of the engram back to Semon.

[21] See Jean-Pierre Dupuy, *Aux Origines des Sciences Cognitives* (1994), trans. M.B. DeBevoise, *The Mechanization of the Mind: On the Origins of Cognitive Science* (Princeton and Oxford: Princeton University Press, 2000). It seems that Byatt read this book in the original in 1994. Its importance for her is indicated by the fact that she refers to it both in 'Fiction informed by science' and in 'Soul Searching'. See A.S. Byatt, 'Soul Searching', *Guardian* 14 February 2004, https://www.theguardian.com/books/2004/feb/14/fiction.philosophy [accessed 28/11.2018].

As pioneered by Norbert Wiener, first-order cybernetics modelled information flow and systems control in mechanical, electronic, and living systems.[22] A founding assumption was that organisms were analogous to machines and that they were regulated according to similar principles of feedback, hierarchical structure, and control: on this basis, Wiener's colleague John von Neumann explored the possibility of developing biological automata and went on to suggest that the gene could be viewed as an information tape that programmed the organism. As living systems were conceptualized as communication systems, the information metaphor began to inflect the way in which genes were understood, as when Watson and Crick argued that the precise sequence of the DNA bases acts as 'the code which carries the genetical information'.[23] Five years later Crick formulated the Central Dogma in terms of a unidirectional flow of information and by 1970 Francois Jacob was able to write that 'heredity is described today in terms of information, messages and code'.[24] Opposition to the information model came primarily from embryologists, who were preoccupied with a question which it could not answer: how, when all the cells in the body have precisely the same genetic information, do they become differentiated during development to become skin cells, liver cells, and so on? Paul Weiss, an embryologist who was present at the Macy conferences, was a key figure in this respect, rejecting the analogy between the organism and the machine and arguing that every cell of the body was made up of processes existing in cooperative interdependence. For Weiss, organisms were self-sustaining systems in which parts and whole were mutually determining, a perspective which anticipates the concept of autopoiesis associated with second-order cybernetics and the work of Humberto Maturana and Francisco Varela.[25]

---

[22] Wiener was a mathematician who developed cybernetics as a formalization of the concept of feedback. His bestseller *The Human Use of Human Beings: Cybernetics and Society* (1950) was instrumental in bringing cybernetics to the wider public.

[23] J.D. Watson and F.H.C. Crick, 'Molecular Structure of Nucleic Acids: A Structure of Deoxyribose Nucleic Acid', *Nature* 17 (25 April 1953), 737–8.

[24] Francois Jacob, *The Logic of Life: A History of Heredity*, trans. Betty E. Spillmann (New York: Vintage Books, 1973), p. 1.

[25] See Humberto R. Maturana and Francisco J. Varela, *Autopoiesis and Cognition: The Realization of the Living* (Dortrecht: D. Reidel, 1980).

Through Marcus, Byatt explores the problematic nature of the application of cybernetic models to biological processes. His PhD project is
on the computer as a model for brain activity but he is uneasy about the
direction the work is taking. His concerns are brought into focus when
a philosopher-colleague, Vincent Hodgkiss, asks the pertinent question
of whether 'the word *information* means the same in all cases, that of
immunology, that of the DNA, that of the mind of the scientist building
a computer'. The question draws attention to the potential disjunction
between information in the cybernetic sense and genetic information as
a component of a living system. Responding to Hodgkiss's comment,
the physiologist Lyon Bowman alludes to Jean-Pierre Changeux's work
on developmental plasticity and suggests that it might offer Marcus a
more productive way into the understanding of brain activity. Byatt
knew of Changeux's work through his contribution to the Piaget–
Chomsky debate, in a paper which was singled out by the editor as
offering a way of reconciling innatist and constructionist positions.[26]
Changeux's argument was that what the genes dictate is a 'genetic
envelope' within which there is room for a degree of plasticity in early
life, as synapses are either stabilized or regress in a kind of adaptive
selection process. As Bowman puts it, he had shown that 'there are
physiological changes—very rapid ones—on growing brains—which
later cease to happen. I should look *there*—' (251). Bowman's intervention prompts Marcus to have a new 'sense of the shape of what he wants
to know' and he subsequently rejects computational models of the mind
and begins to focus on embryology and on phyllotaxis, the development
of patterns on the skin and fur of plants and animals.

Perhaps the most surprising critique of the cybernetic turn comes
from John Ottakar, initially a student and then a lover of Frederica
Potter. Ottakar is a computer engineer who designs shipping systems,
applying the principles of cybernetics in a commercial context.[27]

---

[26] Byatt also draws extensively on Changeux's work for a reading of Donne's poetry.
See A.S. Byatt 'Feeling Thought: Donne and the Embodied Mind' in Achsah Guibbory,
ed., *The Cambridge Companion to John Donne*, pp. 247–58.

[27] While Ottakar the computer programmer laments the loss of humanist values, the
literary Frederica is energized by the new ways of thinking opened up by genetics and
cybernetics. As she comments in *The Whistling Woman* 'explaining genes, and chromosomes, and the language of the DNA [...] it's interesting, it's *what is*' (p. 411, Byatt's italics).
The novels endorse Frederica's perspective, foregrounding the excitement she derives from

His insider's critique is articulated through a paper which he gives to Frederica's evening class on Kafka's *The Castle*. This opens with a reading of the eponymous castle as an image of freedom which is in reality a scene of linguistic and social chaos, as in the myth of the tower of Babel. Ottakar then seizes on K's experience of hearing a strange buzz on the telephone and interprets this in terms of a transition from natural language (represented by the 'heavenly' sound of children singing) to information (represented as a buzzing which can hurt you). In his reading, this is analogous to the switch from the understanding of DNA as a language, which entails interpretation and response, to the concept of the one-way transmission of genetic information. Ottakar develops the point further in a reading of Kafka's 'In the Penal Colony' which suggests a connection between its central trope, the torture machine which writes on the body of the prisoner, and genetic information. He is particularly struck by the fact that the prisoner is killed by the inscription of a sentence, in both the penal and syntactical sense, which he cannot read but which he feels in his body, an insight which signals the epistemic violence entailed in the application of machinic models to living systems. In this respect, Ottakar's description of *The Castle* as 'an inhuman book about being human. Or a human book about being inhuman' also draws attention to the part played by the information metaphor in the emergence of post-humanism. As N. Katherine Hayles has argued, the discourse of information in molecular biology contributed to the belief, which was widespread in the 1990s, that the body could be understood as an information pattern. Information was conceptualized as a kind of 'bodiless fluid' which could flow between different substrates without changing its nature or meaning, making it possible to equate humans and computers and to view natural embodiment as 'an accident of history rather than an

---

thinking about these issues. In this respect, there is a suggestive link between Byatt's fiction and Jacques Derrida's exploration of the philosophical implications of genetics. As Francesco Vitale has shown, in *Of Grammatology* and *La Vie La Mort* Derrida takes the concept of genetic information and fuses it with deconstruction to make the case that *différance* is the 'genetico-structural condition of the life of the living'. See Francesco Vitale, 'The Text and the Living: Jacques Derrida between Biology and Deconstruction', *The Oxford Literary Review* 36 (2014): 95–114, doi: 10.3366/olr.2014.0089 ((101).

inevitability of life'.[28] Through Marcus and John Ottakar, Byatt expresses scepticism about the construal of identity in terms of disembodied information and while not defending the liberal humanist conception of the subject, resists the Cartesian dualism of post-humanism.[29]

Frederica's relationship with Ottakar also plays a part in Byatt's exploration of neo-Darwinian ideas about the evolution of sex, an exploration which extends through this novel and on into *The Whistling Woman*. In the 1960s geneticists were asking why sexual reproduction had evolved at all given that asexual reproduction, through parthenogenesis or budding, is a more efficient process. As Byatt notes in her *Nature* article, this was a question which was occupying John Maynard Smith at the time when *Babel Tower* is set (296). Like George C. Williams, Maynard Smith concluded that sexual reproduction ensured genetic variation, which in turn allowed species to adapt to environmental challenges: genetically identical species, lacking this resource, tended to become extinct over the longer term.[30] In addition, it was thought that sex enabled harmful genetic mutations to be purged, as they would be passed on to some descendants but not others. These ideas are taken up by Luk Lysgaard-Peacock, who is working on two slug populations, one of which is self-fertilizing and genetically identical while the other is sexually propagated and genetically diverse. Subsequently he discovers that the slugs which reproduce parthenogenetically live harmoniously whereas the others fight to the death, tearing each other into a soup, with one or two fit survivors. To interpret these findings, Luk turns to the argument which as we have seen was foundational for both sociobiology and evolutionary psychology, that gendered differences in behaviour derived from the differential reproductive investments of males and females. According to this

---

[28] N. Katherine Hayles, *How We Became Posthuman: Virtual Bodes in Cybernetics, Literature, and Informatics* (Chicago and London: Chicago University Press, 1999), p. 2.

[29] In 'Soul Searching', however, Byatt agrees with Ian Hacking's contention that the twenty-first century is ushering in a new Cartesian phase, as the advent of body parts and prostheses of various kinds is altering our sense of the relationship between minds and bodies. As she puts it, 'the more we control and mess with the wet stuff and with prostheses, the more difficult it is to feel the controlling intelligence as simply part of flesh, blood and wet stuff. (And silicone, possibly).'

[30] See John Maynard Smith, *The Evolution of Sex* (Cambridge: Cambridge University Press, 1978) and George C. Williams, *Sex and Evolution* (Princeton and Oxford: Princeton University Press, 1975).

logic, as males produce large quantities of sperm their interests are best served by competing for and inseminating as many females as possible. Conversely, because eggs are costly to produce, females must select a mate with 'good' genes who will help to raise the offspring, hence female sexual 'coyness'. Due to these conflicts of genetic interest, sex is, as Steve Jones puts it, 'filled with strife'; in this respect it is telling that both Jones's *The Language of the Genes* and Dawkins's *Selfish Gene* have chapters headed the 'Battle of the Sexes'.[31]

Like Maynard Smith, Byatt approaches these issues through a consideration of human parthenogenesis, the as-yet hypothetical alternative to sexual reproduction. John Ottakar is an identical twin and he and his brother understand their relationship in terms of parthenogenesis: we learn from Paul's psychiatrist that he has 'read somewhere that identical twins are a form of virgin birth, of asexual reproduction' and that he is convinced that he is 'the "bud" which formed on the primary zygote which is John'.[32] From Frederica's perspective, this sense of their relationship prompts a failure to differentiate between self and other which is more usually associated with the feminine. This 'feminization' of the twins may reflect the fact that parthenogenesis usually entails female self-replication and for this reason often features in speculative fiction as an image of female emancipation.[33] It takes on this connotation when it crops up in a TV debate between Frederica, Julia Corbett, a novelist who appeared in Byatt's earlier novel *The Game*, and Penny Kumoves, a journalist who specializes in 'the new anxieties of female

---

[31] See *The Selfish Gene*, pp. 140–65 and Steve Jones, *The Language of the Genes* (London: Flamingo, 1994), pp. 96–117.

[32] A.S. Byatt, *A Whistling Woman* (2002) (London: Vintage, 2003), p. 59.

[33] See for example Charlotte Perkins Gilman's *Herland*, published in 1915 during the first wave of feminism, and Naomi Mitchison's *Solution Three*, published in 1973 during the second wave. The association of parthenogenesis with female emancipation also surfaces in *Still Life* in a discussion between Frederica and Raphael Faber. Faber mentions the self-replicating Indian fig in *Paradise Lost* and startles Frederica by identifying it as the Tree of Error which 'produces her daughters from herself like sin and the hellhounds' (p. 257). In this he builds on Alastair Fowler's suggestion in his annotations to *Paradise Lost*: see John Milton, *Paradise Lost* (1968), ed. Alastair Fowler (London: Longman Annotated English Poets, 2nd edition, 2006) p. 534. His comments suggest anxiety about the threat posed by independent women, in a period on the cusp of the second wave of feminism when young women like Frederica were entering higher education in increasing numbers.

graduates' (140).[34] The programme is called 'Free Women', in a direct allusion to Doris Lessing's proto-feminist *The Golden Notebook*, and the discussion focuses on what is perceived as an increasing tendency for women to live with children but without men.[35] In a bathetic echo of Freud's theme of the three caskets, the contributors are confronted with three Tupperware bowls which figure the choices available to women at this time: sex without children; sex and children; and sexual abstinence.[36] Dissatisfied with these alternatives, the contributors seize on the idea that ice applied to the ovaries could in certain conditions produce parthenogenesis: if this were true, women would for the first time have genuine reproductive freedom.

In contrast to this speculative fantasy, the novel maps the existing 'battle of the sexes' through Frederica's marriage to Nigel Reivers, who may know nothing of evolutionary biology but whose behaviour conforms to the pattern outlined by figures such as Maynard Smith and George C. Williams, predicated on the twin concepts of male competition for females and greater female investment in reproduction. In this respect, it is perhaps no coincidence that he is an upper-class landowner whose wealth has come to him through primogeniture, a traditional mechanism for fusing biological and economic forms of male inheritance.[37] Moreover as the narrator explains, he has chosen Frederica for the mother of his child and once their son is born expects

---

[34] The anxieties of female graduates were a hot topic in this period, as evidenced by such studies as Hannah Gavron's *The Captive Wife: Conflicts of Housebound Mothers* (London: Routledge and Kegan Paul Ltd, 1966) and Alva Myrdal and Viola Klein's *Women's Two Roles: Home and Work* (1956), second revised edition (London: Routledge and Kegan Paul Ltd, 1968).

[35] Free Women is the title of the frame narrative of *The Golden Notebook*, which as Lessing says in her Preface to the 1972 edition, was ahead of its time in relation to the Women's Liberation movement. See Doris Lessing, *The Golden Notebook* (1962) (London: Panther, 1972), pp. 9–10.

[36] In Freud's account of the theme of the three caskets the caskets represent women and the third woman signifies death transfigured into beauty by desire: Byatt is making the point that abstinence can be a kind of living death. See Freud, Sigmund, 'The Theme of the Three Caskets', in The Penguin Freud Vol. 14, *Art and Literature* (Harmondsworth: Penguin, 1990), pp. 234–47.

[37] Frederica reflects on the patriarchal structure of Nigel's household in *Babel Tower* (he lives with his wife, his son, and his two older sisters): 'Nigel is like a Pasha in his palace, she thinks, but cannot say. Leo is a male child in a harem' (p. 96).

her to devote herself to him. To this end, he more-or-less imprisons Frederica, preventing her from working, refusing to let her have any contact with her male friends and attacking her with an axe when she attempts to leave. He jealously guards the purity of his genetic inheritance, imposing fidelity on Frederica while at the same time visiting brothels which cater to his specific sexual tastes. As their later divorce hearing makes clear, Nigel's expectations are a throwback to an earlier era of female subjugation and by juxtaposing his behaviour with neo-Darwinian theories of sex, the novel draws attention to the imbrication of these theories with the androcentric and outdated assumptions of the biologists who created them. In this respect, it resonates with feminist critiques of evolutionary biology which were beginning to emerge at the time when it is set. The most prominent figure in this respect was Ruth Hubbard, the first woman to become a tenured lecturer in biology at Harvard before she turned her attention to feminist theory. Hubbard took aim at the naturalization of male competition in the work of Williams, W.D. Hamilton, and others and argued that such explanations of sexual behaviour had been created by privileged men with personal and political interests in describing women 'in ways that make it appear "natural" for us to fulfil roles that are important for their well-being'.[38]

The gendered dynamics of neo-Darwinism are also tested through Luk Lysgaard-Peacock's theory of the redundant male, which anticipates some of the arguments advanced by Jeremy Cherfas and John Gribbin in their 1984 book of that title and by Steve Jones in *The Language of the Genes* and in *Y: The Descent of Men*.[39] Luk is interested in an idea which directly counters the neo-Darwinian axiom discussed above, arguing that the biological costs of sex may be higher for men than women. In a lecture on this subject he emphasizes the elaborate nature of male courting displays, alluding to Darwin's difficulty in accounting for the peacock's tail, and stresses the link between sexual

---

[38] Ruth Hubbard, *The Politics of Women's Biology* (New Brunswick, NJ: Rutgers University Press, 1990), p. 119. See also Anne Fausto-Sterling, *Myths of Gender: Biological Theories about Men and Women* (New York: Basic Books, second revised edition, 1992).

[39] See Jeremy Cherfas and John Gribbin, *The Redundant Male: Is Sex Irrelevant in the Modern World?* (London, Sydney, Toronto: The Bodley Head, 1984) and Steve Jones, *Y: The Descent of Men* (London: Little, Brown, 2002).

reproduction and intense male aggression.[40] His point is that the costs of male dominance and display may be linked to higher rates of accident, illness, and death, in which case men could be considered the 'ultimate losers' in the reproductive struggle. Luk is aware of the problem, wryly noted by Steve Jones, that biologists are often 'painfully close' to the experience they study, and the novel e stresses the potential entanglement of his arguments and his personal history.[41] Luk's relationships follow a pattern which is far from the neo-Darwinian model of male dominance; rather they reflect women's increased agency in relation to sex and reproduction. For example, when Jacqueline Winspear becomes pregnant by Luk, it is she who decides that the pregnancy should continue, and when she miscarries, that the relationship is over; similarly, when Frederica becomes pregnant it is she who takes responsibility for deciding to keep the child. Luk is marginalized in relationship to decisions made possible by increased economic independence for women and improved access to abortion and the pill. By placing Luk's theories in the context of this changed landscape, Byatt makes the point that biological theories are inextricably bound up with their social contexts, in this case what Steve Jones calls in *Y: The Decent of Men* 'the advance of womankind' (241).

In this respect *The Whistling Woman* underscores what Carsten Strathausen has called the 'conceptual instability' of evolutionary theory, which from its inception has migrated back and forth between the biological and social realms.[42] As he notes, Darwin's theory was originally inspired by Thomas Malthus's socio-political arguments and many of his followers were keen to transfer his biological theory back into the social realm. Strathausen also argues that it was this conceptual instability which facilitated the rise of eugenic movements in the twentieth century, an issue which Byatt takes up in relation to another

---

[40] He also points out that the peacock's tail was an instance of male display that repelled Darwin, who told Asa Gray that the sight of the feathers in a peacock's tale made him feel sick. For details see his letter to the American botanist Asa Gray of 2 April 1860, Darwin Correspondence Project, https://www.darwinproject.ac.uk/letter/DCP-LETT-2743.xm [accessed 28 November 2018]. As Byatt explains, she named Luk Lysgaard-Peacock before she knew this story ('Fiction informed by science', 296).

[41] Jones, *The Language of the Genes*, p. 111.

[42] Carsten Strathausen, *Bioaesthetics: Making Sense of Life in Science and the Arts* (Minneapolis: University of Minnesota Press, 2017), p. 70.

speaker at Wijnnobel's conference, Theobald Eichenbaum. Eichenbaum closely resembles Konrad Lorenz, as an ethologist who has strong views about animal instincts and a history of links to National Socialism and Nazi eugenics. As Ute Deichmann has shown, Lorenz joined the National Socialist Party in 1938 and subsequently became a member of its Office for Race Policy.[43] During the war he published a series of papers, known as the brown papers, which argued that civilized societies often fell into genetic decay and that a 'deliberate, scientifically founded race policy 'was needed to counter this tendency.[44] Byatt's protagonist has also published a brown paper, but by emphasizing that this is based on Francis Galton's theories in *Inquiries into Human Faculty*, the novel draws attention to the uncomfortable connection between Nazi eugenics and the development of eugenic thought in Britain. Galton was the founder of eugenics, which he characterized as 'the science of improving stock', and as Byatt's narrator's account of the relevant chapter in his *Inquiries* suggests, there is a clear continuity between his ideas and those of Eichenbaum: for both, the mass of the population is seen as degenerate but capable of uplift through selective breeding.[45] Unsurprisingly, there is a debate over whether Eichenbaum should be allowed to address the conference but in the event he is overwhelmed by his opponents and physically attacked, in an echo of Lorenz's theories about animals banding together ('mobbing') to attack a potential predator. For Lorenz, such behaviour was also present in humans but could be brought under the control of the uniquely human faculty of reason, a belief which the events of the novel seriously call in question.

Indeed, the fact that the conference descends into chaos points to the limits of reason in the form of Wijnnobel's positivist science and his dream of forging 'a biological-cognitive Theory of Everything'. In this ambition he closely resembles E.O. Wilson, who while he acknowledges the importance of the arts and humanities, argues that human nature can 'be ultimately understood only with the aid of the scientific

---

[43] Ute Deichmann, *Biologists under Hitler*, trans. Thomas Dunlop (Cambridge, MA: Harvard University Press, 1996), p. 185.

[44] The papers were known as the brown papers in an allusion to the brown shirts worn by the Nazis.

[45] Francis Galton, *Inquiries into Human Faculty and its Development* (1883) (London: J.M. Dent & Sons), 1943, p. 17. Further references will be given within the main text.

method'.[46] Wilson's *Consilience* was published in the period when Byatt was writing *A Whistling Woman* and the controversy it generated is part of the wider context of her fictional conference. Responding to Wilson's arguments, Stephen Jay Gould contended that scientific methodology was unable to address the central questions of the humanities, and that in consequence the arts and the sciences should be viewed as 'Non-Overlapping Magisteria' which engage in 'intense dialogue' accompanied by 'respectful non-interference'.[47] Byatt's position is closer to that of Gould than Wilson, as is suggested in a discussion with Luk in which Frederica makes the point that as they come from different intellectual traditions, they ask questions which focus on different epistemic objects: whereas Luk wants to know whether the peacock's tail is adaptive, she wants to know why humans find peacock feathers beautiful. However, in the concluding scene of the novel, and hence of the *Quartet* as a whole, Byatt seeks to align these adjacent fields more closely. Frederica has come to tell Luk about their child and as she runs towards him, she reflects on the literature which has shaped her: 'She found herself thinking about *Paradise Lost*, which seemed to float beside her mind like a great closed balloon of its own colour of light, a closed world, made of language, and religion, and science, the science of a universe of concentric spheres which had never existed.' The counterpart to the 'closed balloon' of her consciousness is Luk's 'world of curiosity', while the unborn child between them is 'another person, contained in a balloon of fluid' (420). These nonoverlapping spheres are united by a materialistic monism as Frederica suggests that body and mind, science and art are ultimately composed of one kind of physical stuff: 'She thought that somewhere—in the science which had made Vermeer's painted spherical waterdrops, in the humming looms of neurones which connected to make metaphors, all this was one' (421).[48] Yet the novel ends on a disconcerting note, as the protagonists

---

[46] E.O. Wilson, *On Human Nature* (Cambridge, MA: Harvard University Press, 2004), p. x.

[47] Stephen Jay Gould, *Rocks of Ages: Science and Religion in the Fullness of Life* (London: Vintage, 2002), p. 6.

[48] Byatt's monism is also evident in her appreciative review of Antonio Damasio's *Looking for Spinoza*. See A.S. Byatt, 'The Feeling Brain', *Prospect Magazine*, 20 June 2003, https://www.prospectmagazine.co.uk/magazine/thefeelingbrain [accessed 30 November 2018]. Frederica has been looking at Vermeer's *View of Delft* and has been struck by

gaze at the man-made nuclear early warning system on the North York Moors, described as 'three perfect, pale, immense spheres, like visitors from another planet, angelic or daemonic' (421). These spheres not only mirror the disparate worlds of Frederica, Luk, and their child but also signal the potential and the limits of human intelligence, which has created technologies which may eventually destroy the planet.

Byatt pursues this question further in *The Biographer's Tale*, which explores the prehistory of molecular neo-Darwinism and engages with the ecological issues which were becoming a matter of urgent concern around the time when it was published at the turn of the millennium. The narrative is framed as the biography of a biographer, Scholes Destry-Scholes, who is himself working on Carl Linnaeus, Francis Galton, and Henrik Ibsen. By degrees, the novel's concern with biography turns into a reflection on autobiography, as the frame narrator, Phineas G. Nanson, realizes to his dismay that his account of the biographer and his subjects has turned into 'a first-person story proper, an autobiography'.[49] This prompts him to ask if all writing, including scientific writing, is autobiographical to some degree, which in turn leads him to question the relationship between the scientific observer and their object of study. Foucault broaches this issue in *The Hermeneutics of the Subject* as he explores historically different modes of what he calls 'objectification' and 'subjectification' and it is considered in detail in Lorraine Daston and Peter Galison's analysis of the history of objectivity.[50] As they point out, although we think of objectivity as integral to the scientific method, it did not emerge as a scientific ideal until the mid-nineteenth century, when scientists began to yearn for 'knowledge that bears no trace of the knower—knowledge unmarked by prejudice or skill, fantasy or judgement, wishing or striving'.[51] Prior to this the scientific standard was 'truth to nature', which entailed selecting, comparing and generalizing to arrive at the 'essential'

---

the combination of artistic vision and geometrical and chemical analysis which went into the making of the painting—art and science conjoined.

[49] A.S. Byatt, *The Biographer's Tale* (2000) (London; Vintage, 2001), p. 250.

[50] See Michel Foucault, *Herméneutique du Sujet* (2001), trans. Graham Burchell, *The Hermeneutics of the Subject: Lectures at the Collège de France* (New York: Picador, 2005).

[51] Lorraine Daston and Peter Galison, *Objectivity* (Brooklyn, NY: Zone Books, 2007), p. 17.

phenomenon, in a model which relied on, rather than discounting, the experience and skills of the observer.

The novel maps the transition from truth to nature to objectivity through its exploration of the contrasting worldviews of Linnaeus and Galton. Nanson reads Linnaeus through the lens of *The Order of Things* and sees him as exemplary of the Classical *episteme*, arguing that his taxonomy was designed 'to order and to name the world'.[52] He stresses that Linnaeus's decision to base his taxonomy on the sexuality of plants was arbitrary and that it was an abstract system grounded in numerical calculations but points out that Linnaeus's descriptions of plants are often deeply anthropomorphic, as when he describes *Andromeda poliofolia* as a plant whose beauty is 'preserved only so long as she remains a virgin (as often happens with women also)—i.e. until she is fertilised' (114). Linnaeus also identifies sympathetically with animals and for him the boundary between human and nonhuman is blurred, so that in his *Systema natura* 'the tailed man, *Homo caudatus*, the pygmy and the satyr, which is also the orang-outan' come into the category of anthropomorphs (40). He makes no clear distinction between science and magic, believes that he has a spirit-double, and has an out-of-body experience in which he sees himself lying on a reindeer skin, 'a very corpse, white and cold with staring eyes' while his double journeys to see the Maelstrom, the whirlpool which signifies in its beauty and terror the animating principle of life itself. From the various documents relating to him, Byatt creates a portrait of a scientist from a period before objectivity emerged as a regulatory ideal, for whom the subjectivity and predilections of the observer were necessarily implicated in the construction of the object of knowledge.

In contrast, Galton exemplifies the scientist from Foucault's Modern *episteme* who is committed to the ideal of objectivity. As Daston and Galison have shown, in the mid-nineteenth century scientists 'began to fret openly about a new kind of obstacle to knowledge: themselves' and subjectivity was reconceived as a 'kind of willful self' which endangered scientific knowledge (34–7). The pursuit of knowledge entailed self-

---

[52] In her acknowledgements Byatt points to the significance of Foucault's work for this novel, as well as the *Quartet*. She writes 'I suspect the germ of the novel lies long ago in my own first reading of Foucault's remarks on Linnaeus and taxonomy in *Les mots et les choses*' (*The Biographer's Tale*, p. 264).

discipline and self-denial, a technology of the scientific self, to borrow Foucault's term, which Galton took to extremes, experimenting on himself to the point where on more than one occasion he almost died. His writings reveal that he was haunted by visions of violence and dismemberment although he strenuously denied that they bore any relation to his own 'subconscious' (77). He was also a key figure in what Ian Hacking calls the 'taming of chance' through the development of laws of statistical probability.[53] Galton was among the first to realize that patterns of distribution, such as regression and correlation, could be regarded as autonomous statistical laws, analogous to the laws of nature. However, when applied to people they encouraged a way of thinking about them in terms of normalcy and deviation from the norm and as Hacking points out, 'most of the law-like regularities were first perceived in terms of deviancy: suicide, crime, vagrancy, madness, prostitution, disease'.[54] This led Galton to propose that 'the race' (a term which he used interchangeably for humanity in general and specific groups) could be improved by encouraging breeding among those who conformed 'most nearly to the central type' (10). His famous composite photographs, one of which is reproduced in Byatt's text, capture the process of clustering around the mean in visual terms, the central image representing the shared traits of a given group, the blurred outline representing peripheral outliers. As this suggests, Galton's passion for classification played its part in a wider move to think of people in terms of 'definite classes defined by definite properties': the twenty-first-century drive to geneticize such categories (as in the search for the 'gene for' autism, for example) is not new but can be traced back to Galton's methodologies and preoccupations.[55]

---

[53] Byatt writes appreciatively of *The Taming of Chance* in her essay 'Ancestors', describing it as 'a book which has recently fascinated me, both as a reader, as a writer, and as a human being'. She uses Hacking's argument to develop an illuminating reading of Penelope Fitzgerald's *The Gate of Angels*, suggesting that the novel can be read as a defence of the concept of chance, 'which gives an importance to the individual life, the particular feeling'. See A.S. Byatt, *On Histories and Stories: Selected Essays* (London: Chatto & Windus, 2000), pp. 65–90 (pp. 86, 89).

[54] Ian Hacking, *The Taming of Chance* (Cambridge: Cambridge University Press, 1990), pp. 1,3.

[55] Ian Hacking, 'Making Up People', *London Review of Books* 28: 16, 17 Aug. 2006, https://www.lrb.co.uk/v28/n16/ian-hacking/making-up-people, [accessed 4 November 2018].

Byatt's point is that rather than being a self-evident epistemic virtue, objectivity is a contingent ideal and one which depends on a detachment from the first-person point of view which can be taken too far, leading to a false reductionism. As she makes clear, the question of the value and the limitations of objectivity is fundamental to the arts as well as the sciences, an issue which is taken up in the figure of Henrik Ibsen, who believes that his work depends on access to an inner self which he bears deep within him and which he must protect at all costs from the incursions of others. At the same time, he contends that the author's task is 'to see, not to mirror' and claims that as a dramatist he has had to 'kill or drown [my] own personality', while his aesthetic is further complicated by the fact that, as his biographer points out, drama is a collaborative art and in making characters real, the author must leave that reality sketched and incomplete, to be fleshed out 'not by one, definitive, magisterial actor, but a succession of these too fleshy ghosts each filling out different pouches and pockets' (84). Like many nineteenth-century realists, Ibsen is caught in a dialectic between objectivity and subjectivity, perspectives which, as Thomas Nagel argues in *The View from Nowhere*, are not opposed; rather, the distinction between subjective and objective views is a matter of degree across a wide spectrum. However, as Nagel also suggests, these perspectives are 'uneasily related' and may be irreconcilable, a point Byatt makes through a description of Ibsen's discrepant eyes: 'one was large, I might almost say horrible—so it appeared to me—and deeply mystical; the other much smaller, rather pinched up, cold and clear and calmly probing' (80).[56]

The novel also speaks to its millennial moment through the frame narrative in which Nanson reflects on his trajectory from postmodern literary critic to biographer to parataxonomist, a trajectory which reflects that of writers like Byatt herself who were disillusioned with postmodernism and liberal humanism and eager to engage with biological perspectives on human nature. The specific trigger for Nanson's turn to parataxonomy is that one of his lovers, Fulla Biefield, is a bee taxonomist and paleo-ecologist. When they first meet Fulla is scathing about his response to *Silent Spring*—he has written a paper linking it to

[56] Thomas Nagel, *The View from Nowhere* (1986) (Oxford: Oxford University Press, 1989), p. 4.

'popular-cultural images of induced panic'—and stresses the urgency of a situation where humans are destroying 'every day 6,000 species perhaps, many unknown, some perhaps essential—certainly essential—to the survival of a whole chain of others'. In common with figures like Daniel Janzen (who pioneered the practice of para-taxonomy), she is concerned with the reciprocities between organisms which have evolved over millions of years but which are now under threat due to human activity.[57] As a taxonomist she is aware that the critical difference between her perspective and that of Linnaeus is that 'he didn't see—he didn't need to see—the interdependence of things' (121). For her, the advent of the Anthropocene has prompted a reconceptualization of nature in terms of the concepts of mutualism and coevolution. She is also aware that the Human Genome Project is likely to reveal more about the close genetic relationships between humans and other organisms, as 'the new genetic groupings—the clades—are going to sweep away the Linnaean families and genera and species, and reconnect everything quite differently. It is possible, for example, that a mushroom is more nearly related to you than to a chrysanthemum or a slime-mould'. Her comments anticipate some of the unexpected outcomes of the Human Genome Project, for example, the finding that humans share 92 per cent of their DNA with mice.

While Fulla is a harbinger of the future, Nanson sees himself as part of the 'great tradition' of evolutionary biology, which for him is exemplified by the work of W.D. Hamilton, one of the key architects of neo-Darwinism. Like Galton, Hamilton deploys statistical methods to make claims about the distribution of traits, claims which are predicated on genes being discrete units of inheritance with a direct relationship between genotype and phenotype. Hamilton's argument for the genetic basis of altruism was extraordinarily influential, especially after its popularization in *The Selfish Gene*, but in his essays Hamilton expresses unease about the threat the theory posed to traditional views of human nature, confessing his 'dislike for the idea that my own behaviour or behaviour [sic] of my friends illustrates my own theory of sociality or

---

[57] Daniel Janzen is an evolutionary ecologist who wanted to inventory the biodiversity of Costa Rica, which lacked the resources to train large numbers of graduates. Accordingly, he trained members of the rural population to carry out the work in collaboration with specialist taxonomists.

any other. I like always to imagine that I and we are above all that, subject to more mysterious laws.'[58] As this suggests, his essays are an example of the neglected genre of scientific life-writing, in which autobiography is interwoven with scientific exposition to construct a picture of the scientific self which includes aspirations and beliefs which are erased in formal scientific publications (217).[59] Like the self which is articulated in any autobiography, such a scientific self is a construction but one which illuminates the complex interplay between scientific theories and their broader social and historical context. In this respect, it could be argued that the novel itself gradually morphs into a form of scientific life-writing, as Nanson reflects on the significance of his new profession of taxonomy. In the context of the Anthropocene, this requires him to abandon human exceptionalism and to see humans as one strand within a complex web of interdependencies. Accordingly, he observes stag-beetles from a perspective of 'critical anthropomorphism', a perspective grounded in the belief that we can learn a great deal about other organisms through empathetic identification as we share so many of their emotional and cognitive processes.[60] He reads the behaviour of stag beetles through the lens of his own experience of male aggression and endorses Hamilton's view of competition as matching 'my own sense of the nature of things' (253).

The Biographer's Tale returns to the question of the analytic of finitude as Nanson distinguishes between 'human truths' and truths about the world, arguing that:

---

[58] W.D. Hamilton, *Narrow Roads of Gene Land: The Collected Papers of W.D. Hamilton, Vol. 1, Evolution of Social Behaviour* (1996) (Oxford: Oxford University Press, 1998), p. 2.

[59] Despite the considerable critical attention which has been given to forms of life-writing in the last two decades, scientific life-writing has generally been neglected by literary scholars. However, some of the relevant issues are addressed in Sam Ferguson, 'Why Does Life-Writing Talk About Science? Foucault, Rousseau and the Early Journal Intime (*Biography* 40 (Spring 2017), 307–22.

[60] Critical anthropomorphism was pioneered by Gordon Burghardt in the 1980s and has subsequently been championed by the primatologist Frans de Waal. For an illuminating discussion of the issues see Frans de Waal, 'Anthropomorphism and Anthropodenial: Consistency in Our Thinking about Humans and Other Animals', *Philosophical Topics* 27 (Spring 1999), 255–80. Byatt has written about her 'instinctive aversion' to anthropomorphism but acknowledges that the discovery of 'the great extent to which DNA patterns are shared by all creatures' has changed her view of commonalities in the natural world: see *On Histories and Stories*, p. 80.

There are very few truths about the world, but the truth about *that* is that we don't know what we are not biologically fitted to know, it may be full of all sorts of shining and tearing things, geometries, chemistries, physics we have no access to and never can have. Reading and writing extend—not infinitely, but violently, but giddily—the variations we can perceive on the truths we thus discover. Children are afraid of the dark; a double walks at our side, or hangs from our flesh like a shadow; and we put a whole life-time (which is brief indeed in the light *even of history* let alone of the time of the world) to discovering what these things mean for us—dark, and shadows. (237)

Like Wijnnobel in the *Quartet*, Nanson underscores man's finitude, that is, the fact that his knowledge is constrained by what Foucault calls its 'anatomo-physiological conditions'.[61] Moreover, he endorses Foucault's view that in the modern *episteme* man is construed as an empirical being who must at the same time be the source of the representations through which he knows the empirical world: in this respect he is what Foucault calls an 'empirico-transcendental doublet'. As we have seen, one way of dealing with this aporetic state of being is to espouse a positivism which effectively reduces the transcendental to the empirical. Rather than taking this line, Nanson suggests that literature allows us to map the self-division arising from our status as both subjects and objects of knowledge and to probe the limits of human consciousness and temporality. As Nanson points out, all the inset stories in *The Biographer's Tale* concern 'ghosts and spirits, doubles and hauntings, metamorphoses, dismemberment, death' and as such they map the haunted landscape of human finitude (237).

Byatt argues in the essay 'Ancestors' that fictions take their form from 'the large paradigmatic narratives we inhabit' and that the most significant shift of the last two thousand years has been from the biblical narrative of 'creation, salvation and resurrection' to the

---

[61] Nanson's perspective can be aligned with the correlationism which has been critiqued by speculative realist philosophers, most notably Quentin Meillassoux, who defines correlationism as the belief that we can only know the correlations between thought and being and not things in themselves. As he summarizes it, the correlationist position is that 'thought cannot get *outside itself*' (Quentin Meillassoux, *After Finitude: An Essay on the Necessity of Contingency*, quoted in Steven Shaviro, *The Universe of Things: On Speculative Realism*, Minneapolis: University of Minnesota Press, 2014), p. 134.

Darwinian narrative of human origins, a transition given added impetus by neo-Darwinism.[62] While Byatt is no hard-line determinist, she takes from the neo-Darwinian framework the belief that identity crystallizes around an unalterable genetic core. Her fiction thus adheres to the 'crystal' rather than the 'flame' model of living systems, a point which has been obscured by comments in her *Nature* article which appear to conflate the two.[63] However, in her fiction, Byatt acknowledges their complex history and ontological implications. In *Babel Tower*, for example, a biochemist refers to the origin of these ideas in Edwin Schroedinger's intuition that genes were aperiodic crystals, an intuition which initiated two very different research paths. On the one hand, it led to the fusion between cybernetics and molecular biology discussed earlier in this chapter, in which life was reconfigured as an information system, a perspective exemplified when Byatt's scientist suggests that 'the *whole universe* might be an information system—of messages flowing through crystals amongst parasitic noise' (251). In this view, order is imposed on disorder due to the existence of what Piattelli-Palmarini calls an underlying plan, 'distinct from its material expressions and containing from the very outset the "envelope" of its possible manifestations', a genetic programme which predetermines the 'geometry' of the organism.[64] The other research path turns on the concept of autopoiesis, that is, the idea that in living systems as well as the physical universe, order at the macro level arises from disorder at the micro level. From this perspective, the origin of life lies in self-organization, which must precede, logically, the development of a programme. In its elaboration by Maturana and Varela in the 1980s, this theory suggests that living organisms 'become real and specify themselves' through their autopoietic organization, in which the parts and the whole are mutually determining.[65]

---

[62] A.S. Byatt, *On Histories and Stories*, pp. 65–6.

[63] Byatt writes that the concept of order from noise 'can be made into metaphor as the order of the shape of a flame as it burns or of a crystal as it grows' but does not mention that, as Piatelli-Palmarini explains, the two metaphors denote very different scientific and ontological commitments. See Byatt, 'Fiction informed by Science', 296.

[64] Massimo Piaittelli-Palmarini, *Language and Learning*, p. 8.

[65] See Humberto R. Maturana and Francisco J. Varela, *The Tree of Knowledge: The Biological Roots of Human Understanding* (1987), trans. Robert Paolucci, revised edition (Boston and London: Shambhala, 1992), p. 48. Further references will be given within the main text.

The implications of Byatt's adherence to the crystal model and to the neo-Darwinism associated with it are clarified in her essay 'Faith in Science', which offers an overview of the 'Darwin Wars' provoked by sociology and evolutionary psychology.[66] While Byatt is critical of the overheated rhetoric used by both sides in the debate, she is explicit about her allegiance to neo-Darwinism. She describes her frustration with the 'overmastering systems of explanation' offered by Marxism, Freudianism, and feminism in the later twentieth century and her sense that their constructionist descriptions of the human were 'misleading, or inadequate, or couched in loose, or meaningless, or tautological language'. In contrast, she explains that John Maynard Smith's account of Weissman's barrier enabled her 'to understand something very precise about heredity and mortality', about what an individual was and her relation to others. This endorsement of neo-Darwinism is amplified as she introduces Lionel Trilling's concept of biological reason:

> As long ago as 1955 Lionel Trilling, in a lecture entitled 'Freud: Within and beyond Culture' [sic] praised Freud for offering resistance to 'cultural omnipotence'. Trilling said that Freud described, 'somewhere in the child, somewhere in the adult…a hard, irreducible, stubborn core of biological urgency, and biological necessity, and biological reason, that culture cannot reach and that reserves the right, which sooner or later it will exercise, to judge the culture and resist and revise it.'

In line with Trilling's argument, Byatt's fiction is shaped by the idea of an irreducible biological core, or programme, which is separable from and impervious to culture. Nature and culture are parsed as distinct entities which are also in competition, raising the perennial question of which has the more powerful influence on human behaviour: depending on the answer to this question, the agency of human beings is seen

---

[66] A.S. Byatt, 'Faith in Science', https://www.prospectmagazine.co.uk/magazine/faithinscience [accessed 26 November 2018]. Byatt takes the term 'the Darwin Wars' from the book of that name by Andrew Brown which she mentions in the article. See Andrew Brown, *The Darwin Wars; the Scientific Battle for the Soul of Man* (New York: Pocket Books, 2000). In his account, the warring sides are the neo-Darwinians Richard Dawkins, John Maynard Smith, and Daniel Dennett, while their critics are the 'Gouldians', Stephen Jay Gould, Richard Lewontin, and Steven Rose.

as constrained to a greater or lesser degree. The autopoietic perspective, in contrast, stresses that autopoietic systems engage in 'structural coupling' with the environment so that there is no discontinuity between 'what is social and what is human and their biological roots'. Moreover, from this point of view, the living organism is agentic in the sense that 'it pulls itself up by its own bootstraps and becomes distinct from its environment through its own dynamics' (46–7).[67] This perspective is not available to Byatt's protagonists, who more often understand themselves as subject to biological reason rather than as active participants in it. This is particularly marked in relation to the representation of the reproductive lives of women in the *Quartet*. Byatt is wary of arguments for fixed gender differences and is also alert to the porous boundary between the biological and the social and the fact that biological theories are inflected by the social and cultural contexts in which they are forged. Nonetheless, there is a sense in which the mind/body dichotomy which runs throughout Byatt's fiction is recast in the *Quartet* in terms of the genes versus environment debate.[68] We can see this process at work in *The Whistling Woman* in relation to Frederica's second pregnancy. In the Lessing-inspired 'Free Women' TV debate mentioned above, Frederica argues that 'the body [...] wants to be pregnant. The woman often doesn't', thus dividing the woman, the person with aims and ambitions, from her body (143). When she begins a relationship with Luk Lysgaard-Peacock, it is amidst intense discussion of neo-Darwinian theory and evolutionary explanations of sex, of male display and female choice, so that when she becomes pregnant it is as though her freedom has been compromised by a set of evolved mechanisms that 'want' her to be pregnant: natural selection

---

[67] Maturana and Varela argue that autopoietic systems are closed systems in that their integrity depends on what they called 'operational closure'. However, it is important to stress that the idea of a closed system coexists with the idea of openness, as circular causality is coupled with the system's adaptive dependence on its environment. See David Radrouf, Antoine Luz, Diego Cosmelli, Jean-Philippe Lachaux, Michel le Van Quyen, 'From autopoiesis to neurophenomenology: Francisco Varela's exploration of the biophysics of being', *Biological Research* 36 (2003), 21–59.

[68] Byatt has frequently commented on the importance of the mind/body problem for the female protagonists in her fiction, noting for example that *The Virgin in the Garden* is a novel 'about the body–mind problems of a young woman interested in her own sex—versus intellect conflict, and also in the nature of metaphor'. See Byatt, 'Fiction informed by science', 295.

has done her thinking for her. At the end of the novel, an allusion to *Paradise Lost* ('the world was all before them, it seemed') positions Frederica and Luk as inheriting the fallen world of evolutionary biology, their position reflecting the situation Byatt describes in 'Faith in Science', in which 'biological reason' requires us to redefine fundamental concepts like love and charity. 'We shall think of something', says Luk, but the task of negotiating between biological reason and a residual humanism lies beyond the timeframe of the *Quartet* (421).

| 3 |

Ian McEwan

The Literary Animal

I an McEwan's interest in neo-Darwinian theory is well-documented in interviews and on his website, which archives his essays on the subject and podcasts of debates with figures such as Richard Dawkins and Stephen Pinker. His engagement dates back to the 1990s, a time of growing excitement surrounding the Human Genome Project, which promised to uncover nothing less than the ultimate causes of human behaviour.[1] It was also a decade which saw a number of bestselling books promoting a neo-Darwinian view of human nature, including Steven Pinker's *The Language Instinct*, Matt Ridley's *The Origins of Virtue*, E.O. Wilson's *Consilience*, and Robert Wright's *The Moral Animal*. These books were underpinned by a gene-centric view of evolution and took inspiration from the burgeoning field of evolution-ary psychology, effectively a continuation of the sociobiological project which aimed to uncover the biological roots of human behaviour. For

---

[1] The hubristic ambitions of the Human Genome Project (HGP) are well-captured in Walter Gilbert's essay 'Vision of the Grail' in which he anticipates a future in which 'Three billion bases of sequence can be put on a single compact disc (CD), and one will be able to pull a CD out of one's pocket and say, "Here is a human being; it's me"' (W. Gilbert, 'Vision of the Grail' in D.J. Kevles and L. Hood (eds) *The Code of Codes* (Cambridge: Harvard University Press, 1992), 83–97, quoted in Evelyn Fox Keller, *The Century of the Gene* (Cambridge and London: Harvard University Press, 2000), p. 6.

Genetics and the Literary Imagination. Clare Hanson, Oxford University Press (2020). © Clare Hanson.
DOI: 10.1093/oso/9780198813286.001.0001

its exponents, evolutionary psychology had the potential to transform our understanding of humanity, for as Wright explains it touches on 'just about everything that matters: romance, love, sex...friendship and enmity...selfishness, self-sacrifice, guilt'.[2] The obvious overlap between the interests of evolutionary psychology and those of literature was largely responsible for the phenomenon of 'literary Darwinism', a branch of criticism grounded in the belief that an evolutionary perspective can offer radically new insights into the meaning of literary texts. McEwan contributed to *The Literary Animal*, a collection of essays on literary Darwinism, and his fiction intersects in multiple ways with evolutionary theory and the literary criticism associated with it. This chapter maps the arc of his engagement with evolutionary psychology, which has been more sustained than previously recognized, and reads his changing relationship with the field as an index of shifts in the public understanding of evolutionary biology.

For McEwan, evolutionary theory provides us with 'a creation story of a grandeur and beauty unrivalled by that of, say, Genesis, or the dreaming snakes of the Australian aborigines'.[3] Neo-Darwinism added a model of genetic causation to this creation story, which enabled sociobiology to reconfigure original sin in terms of genetic endowment. This perspective is amplified in evolutionary psychology, whose exponents argue that to ensure their own proliferation, genes forge 'mental modules' in which the rules of human behaviour are encoded.[4] These modules are analogous to computer programs and constitute an innate architecture of the mind, a perspective which directly challenges the 'blank slate' or constructionist view of the mind which, according to evolutionary psychologists, dominated twentieth-century social science. The emphasis on innateness is given a further twist as it is argued that these modules evolved to solve problems which confronted humans in the ancestral past and that in consequence, human behaviour is often better adapted to the Pleistocene era than to the modern

[2] Robert Wright, *The Moral Animal: Why We Are The Way We Are* (London: Abacus, 1996), 5. Further page references will be given within the main text.
[3] Typescript for Radio 4 talk (untitled), c 1998, Box 29.3, 2, Ian McEwan Archive, Harry Ransom Center, University of Texas at Austin.
[4] See L. Cosmides and J. Tooby, 'The Psychological Foundations of Culture', in *The Adapted Mind: Evolutionary Psychology and the Generation of Culture*, eds. Jerome H. Berkow, Leda Cosmides, and John Tooby (New York: Oxford University Press, 1992), pp. 19–136.

world. For example, in the hunter-gatherer past, evolution would have favoured the male pursuit of status because it translated into multiple sexual partners and numerous offspring; despite changes in sexual mores, evolutionary psychologists argue that the biological drive to competition persists, leading to the male status hierarchies which are universal in human society.[5] More broadly, they contend that society is organized around the twin poles of competition and co-operation. Competition is explained by the fact that individuals are proxies for 'selfish genes' but cooperation poses more of a problem, for if selection favours adaptations which increase fitness, there can be no logic in benefitting another person at a cost to oneself. In addressing this issue, evolutionary psychologists drew on William Hamilton's theory that there could be a trade-off between individual fitness and that of genetically related kin, so that in some circumstances it would make sense to sacrifice oneself to further the spread of shared genes. The theory of kin altruism was extremely influential, as was Robert Trivers' concept of reciprocal altruism, whereby altruism towards unrelated individuals is linked to the expectation of reciprocal future benefit, a model supported by game-theoretical analysis. Building on these theories, the prominent evolutionary psychologists Leda Cosmides and John Tooby argue that altruism 'can be as natural as selfishness' in human society.[6] However, critics have argued that this holds true only for a definition which elides altruism's distinguishing feature, that of *disinterested* concern for others.[7] As Wright acknowledges, evolutionary

[5] For an incisive critique of evolutionary psychology's deeply conservative approach to gender roles, see Anne Fausto-Sterling, 'Beyond Difference: Feminism and Evolutionary Psychology' in *Alas Poor Darwin*, eds. Hilary Rose and Steven Rose (London: Vintage, 2001), pp. 74–189. As she suggests, evolutionary psychology's 'hard-wired view of the inflexibility of human social arrangements' flies in the face of the evidence of the plasticity of behaviour. Depending on their environments, both sexes can exhibit a wide range of behaviours (p. 183).

[6] Leda Cosmides and John Tooby, 'Cooperation', in Berkow, Cosmides, and Tooby, p. 161.

[7] The term altruism was coined in 1851 by Auguste Comte to denote the ideal of living for others ('autrui') which was central to his Positivist philosophy. As Suzanne Keen has shown, recent debates over altruism are prefigured by those of the mid-nineteenth century, when the question of whether human beings are capable of disinterested feeling for others was hotly debated. Comte's 'religion of humanity' was attractive to many novelists, most notably George Eliot, whose partner G.H. Lewes introduced Comte's thought to Britain, and his ideas were widely diffused in nineteenth-century literature. The evolutionary theorist Herbert Spencer was more sceptical, however, arguing that egoism is fundamental to human nature and that unselfish concern for others had

psychology is often seen as promoting a 'cynical' view of human relations in which 'natural selection has crafted an ever-expanding web of affection, obligation and trust out of ruthless genetic self-interest' (202).

Evolutionary psychology has been strongly criticized from a range of perspectives. Stephen Jay Gould has offered the most trenchant scientific critique, arguing against its 'Darwinian fundamentalism'. For Gould, this fundamentalism consists in the belief that a single principle, that of natural selection, can account for 'the full complexity' of evolutionary outcomes.[8] He offers three pluralist correctives to this model. The first is his concept of punctuated equilibrium, which complicates Darwin's idea of gradual change over geological time, introducing the idea that most new species originate in a geological 'moment'. The second is the concept of 'spandrels', his term for features which are non-adaptive, but which nonetheless feed into evolutionary change, while the third is the role of contingency and chance in the history of life, a history which has no inherent directionality. Gould also contests the claim that modern human behaviour is governed by adaptations which arose in the Pleistocene, pointing out that we have scant knowledge about the evolutionary past and that such arguments are untestable and unscientific (101). The science commentator Kenan Malik picks up on this point in his *Man, Beast and Zombie* (2002), arguing that that the idea that humans are better adapted to the ancestral past than the modern world appeals to and reflects a specifically millennial sense of anxiety and dislocation. The novelist Marilynne Robinson takes the argument a stage further in her analysis of evolutionary psychology, linking its bleak assessment of human nature to the broader context of post-Enlightenment modernity. For Robinson, what is striking about such accounts of human experience is their exclusion from consideration of 'the felt life of the mind', an exclusion she attributes to a mistaken confidence that scientific knowledge will

---

developed only recently in human beings. See Suzanne Keen, 'Altruism Makes a Space for Empathy', branch collective, http://www.branchcollective.org/?ps_articles=suzanne-keen-altruism-makes-a-space-for-empathy-1852 [accessed 21 March 2018].

[8] Stephen Jay Gould, 'More Things in Heaven and Earth', in *Alas Poor Darwin*, pp. 85–105, (p. 94). Further page references will be given within the main text.

answer 'essential questions about the nature of reality'.[9] Countering this assumption, she highlights the way in which the redefinition of altruism in terms of cost/benefit analysis disallows 'the intense and emotional subjective considerations a human altruist is likely to ponder', substituting for these an 'untestable mathematical formula' (63). For her, evolutionary psychology is an assault on our self-understanding, representing the mind as 'a passive conduit of other purposes than those the mind ascribes to itself' (71).

Enduring Love announces its engagement with these issues in its opening scene, a much-praised description of an accident involving a helium balloon. A man is struggling to anchor the balloon against a backdrop of strong winds but has become caught in the guy ropes while his grandson lies in the basket paralyzed by fear. As a violent gust lifts the balloon into the air, five men run to help, aiming to hang on to the ropes and bring it back to earth. However one by one they drop away, with the exception of one, John Logan, who subsequently falls to his death: the boy is unharmed. The accident and its aftermath are filtered through the first-person narrative of one of the participants, Joe Rose, a science journalist who is profiting from the contemporary publishing vogue for 'every possible slant on neo-Darwinism, evolutionary psychology and genetics'.[10] Given his extensive knowledge of the field it is hardly surprising that he interprets the accident in terms of evolutionary psychology, seeing it as the outcome of a Prisoner's Dilemma (one of the games deployed by Hamilton and Trivers) in which the participants must calculate the relative benefits of competition and cooperation. According to Joe, for the men holding the ropes the scales tipped as soon as one person let go: 'Suddenly the sensible choice was to look out for yourself. The child was not my child and I was not going to die for it. The moment I glimpsed a body fall away—but whose?—and I felt the balloon lurch upwards, the matter was settled; altruism had no place' (15). From the point of view of evolutionary psychology,

---

[9] Marilynne Robinson, *Absence of Mind: the Dispelling of Inwardness from the Modern Myth of the Self* (New Haven and London: Yale University Press, 2010), p. 33. Further page references will be given within the main text.

[10] Ian McEwan, *Enduring Love* (London: Vintage, 2006), p. 70. Further page references will be given within the main text.

'morality' consists solely in maximizing the interests of one's genes, so that Joe has made a rational, unexceptionable choice.

The ramifications of Joe's decision are explored in the light of the competing perspectives of evolutionary psychology, religion, and literature and in this respect the novel can be seen as an intervention in the 1990s 'science wars' triggered by the physicist Alan Sokal's placing of a hoax article in *Social Text*. This purported to give a social constructionist reading of quantum gravity and Sokal's success in placing it in a cultural studies journal prompted him to claim that humanities scholars not only had no understanding of science but lacked a robust critical methodology of their own (he was particularly scathing about postmodern relativism). The incident prompted a good deal of soul-searching among scholars in the humanities who, as John Guillory has shown, struggled to make the case for their interpretive methods as valid epistemologies.[11] In the novel, Joe represents the epistemological confidence of science, or more precisely of scientism, given that he endorses the extension of natural scientific methodologies into the social sciences and the humanities. He is particularly enthusiastic about the potential of evolutionary psychology in this respect but through his characterization the novel points to its potential limitations for parsing human nature. In many respects, Joe resembles the hypothetical subject of evolutionary psychology, navigating his way through the world according to a calculus of rational self-interest. However, he finds it difficult to connect thought with feeling: as Clarissa reflects, 'the trouble with [his] precise and careful mind is that it takes no account of its own emotional field' and in consequence, his decisions are often inappropriate. Clarissa's insight into Joe coheres with Antonio Damasio's argument that emotions are integral to the reasoning process, which was inspired by his observation of brain-damaged patients who retained rational capacity but whose emotional responses were impaired, leading to poor decision-making. While Joe's condition is not pathological, he resembles these patients in his abstract, affectless approach to personal relations.[12]

[11] John Guillory, 'The Sokal Affair and the History of Criticism', *Critical Inquiry* 28 (Winter 2002), pp. 470–508 (p. 501).
[12] Damasio is a background presence throughout *Enduring Love* and his *Descartes Error* is cited in the acknowledgements. In his notes for the novel McEwan comments on

McEwan subjects evolutionary psychology to additional scrutiny through the relationship between Joe and Jed Parry, one of the protagonists in the balloon accident who in its wake develops a form of erotomania focused on Joe. Joe correctly diagnoses this as de Clérambault's syndrome, a rare form of obsession which involves the stalking of the victim and the conviction that one's love is returned. However, it is Jed's religious beliefs which bear most directly on the novel's evolutionary interests. Jed is not affiliated to any church and as Joe comments disdainfully, he is vague on the specifics of doctrine, but he is consistent in opposing Joe's scientism. Accusing Joe of contempt for religion, he argues that his speculations about how life began on earth constitute 'a puny rant against an infinite power'. Yet as the relationship between the two men unfolds, McEwan underscores the fact that in one respect their worldviews mirror each other (135). As Dorothy Nelkin has pointed out, evolutionary psychologists often resemble missionaries, 'advocating a set of principles that define the meaning of life and seeking to convert others to their beliefs' and although their beliefs are not theistic they tend to present the world in terms of 'cosmic principles, ultimate purpose and design'.[13] So in the novel, Jed's invocation of the 'infinite power' of God is matched by Joe's description of the 'mysterious powerful force' which binds hydrogen and oxygen molecules to form water. Joe's choice of words underscores the homology between the idea of divine causation and what Gould identifies as evolutionary psychology's 'fundamentalist' insistence on natural selection as the sole driver of evolution. As Gould notes, this belief—or dogma—persists despite the growing evidence that natural selection cannot explain all aspects of evolution.[14]

Clarissa represents the epistemological claims of literature and the fact that she is a Keats scholar places her in implicit dialogue with

---

Joe's emotional inadequacy in relation to Clarissa: 'the point is that Joe does not know that he was in any fundamental sense wrong, or that C has any significant truth to impart; or that there could be mutual forgiveness on that basis. Tolerance at most.' Green Notebook for *Enduring Love*, September 1995–January 1997, Container 7.10, unnumbered page, Ian McEwan Archive, Harry Ransom Center, University of Texas at Austin.

[13] Dorothy Nelkin, 'Less Selfish than Sacred? Genes and the Religious Impulse in Evolutionary Psychology', in *Alas Poor Darwin*, pp. 14–27 (p. 19).

[14] Gould, 'More Things in Heaven and Earth', pp. 89–90.

Richard Dawkins, whose intervention in the science wars came in the wake of his appointment to the first chair in the Public Understanding of Science at Oxford University.[15] In a series of public lectures which were subsequently collected in *Unweaving the Rainbow*, Dawkins acts as a flag-bearer for the kind of 'third culture' envisaged by his literary agent John Brockman. Rather than mediating between the two cultures, Brockman's third culture is one in which empirical science takes over the traditional territory of the humanities as it explains 'the deeper meaning of our lives'.[16] In line with this project, Dawkins robustly defends the truth of scientific reductionism and challenges what he sees as hostility towards science in literature, thereby positioning himself within a long tradition of debate about science and literature in the UK.[17] He is particularly concerned to make the case for the beauty of reductionist science, taking up Keats's famous claim that Newton's analysis of the rainbow had destroyed its beauty and countering it by arguing that 'mysteries don't lose their poetry because they are solved. Quite the contrary; the solution often turns out more beautiful than the puzzle.'[18] However, not everyone was convinced by these arguments. For example, in a review of *Unweaving the Rainbow*, the philosopher Thomas Nagel argues that there is a difficulty with the idea that the

---

[15] The Simonyi Professorship for the Public Understanding of Science was established in 1995 on the understanding that Dawkins would be its first holder. Although Dawkins is not mentioned in the novel he appears in the draft notes where McEwan writes: 'Did J go to a party the night before picking up Clarissa from the airport? Was Dawkins there— event glamour.' This suggests that Dawkins was in his mind as the most influential advocate of the theories espoused by Joe. Green Notebook for *Enduring Love*, September 1995–January 1997, Container 7.10, unnumbered page, Ian McEwan Archive, Harry Ransom Center, University of Texas at Austin.

[16] John Brockman, 'The Third Culture', https://www.edge.org/conversation/the-emerging [accessed 24 September 2017]. This is the Introduction to his *The Third Culture: Beyond the Scientific Revolution* (New York: Simon and Schuster, 1991). The impact of Brockman's book was augmented by the success of the *Edge* website which he founded and which popularized the work of writers such as Dawkins and Steven Pinker.

[17] See Helen Small, *The Value of the Humanities* (Oxford: Oxford University Press, 2013) for an excellent discussion of literature and science debates. As she suggests, in Britain these reach back through the Snow–Leavis controversy of the 1950s, to the Huxley–Arnold debate of the 1880s, then to Carlyle and Coleridge's opposition to Bentham and James Mill in the nineteenth century.

[18] Richard Dawkins, 'Science, Delusion and the Appetite for Wonder', the Richard Dimbleby Lecture, televised 12 November 1996 on BBC 1. Text from Edge.org, https://www.edge.org/conversation/richard_dawkins-science-delusion-and-the-appetite-for-wonder,%20com [accessed 22 March 2018].

beauty of science and the beauty of poetry are of the same kind. As he suggests, reductionism is associated with 'a very special hunger for understanding, the hope for universal order and reduction of complex variety to simple elements'. Such an impulse is the antithesis of the aesthetic disposition towards complexity and multivalence, leading him to conclude that reductionist science 'is not poetry—it is not like any art—and its effect on us does not require poetic forms of presentation' (as Dawkins claimed that it did).[19]

These issues are taken up in a series of conversations between Joe and Clarissa. In one session Joe, like Dawkins, invokes Keats's response to Newton and argues for the wonders of reductionist science, while Clarissa, rather like the humanities scholars who responded to the Sokal affair, struggles to articulate her understanding of the 'whole' that might be lost in such analysis. However, in a later scene in which her godfather and Joe are discussing the beauty of reductionist explanations, she intervenes in the debate by quoting from Keats's 'Endymion': 'Be still the unimaginable lodge / For solitary thinkings; such as dodge / Conception to the very bourne of heaven, / Then leave the naked brain:'. In these lines, Keats reflects on the difficulty of connecting subjective consciousness ('solitary thinkings') with the physical or neurological '(the naked brain') and in quoting them Clarissa raises the question of what is often seen as the limit case for reductionist explanation, the so-called 'hard problem' of consciousness. In David Chalmers' formulation, this is the difficulty of explaining subjective consciousness in physical terms: as he puts it, 'how can we explain why there is something it is like to entertain a mental image, or to experience an emotion? It is widely agreed that experience arises from a physical basis, but we have no good explanation of why and how it arises.'[20] Supporters of evolutionary psychology such as Daniel Dennett have argued that the hard problem is itself an illusion, a fantasy of the irreducibility of subjective consciousness, but the novel suggests otherwise. Reflecting on the quotation and on the brevity of Keats's life,

[19] Thomas Nagel, 'Why so Cross?', *London Review of Books*, 21: 7, 1 April 1999, pp. 22–3, https://www.lrb.co.uk/v21/n07/thomas-nagel/why-so-cross [accessed 25 September 2017].

[20] David Chalmers, *The Character of Consciousness* (Philosophy of Mind) (Oxford: Oxford University Press, 2010), p. 5.

Joe falls into a 'daydream' in which for once he gives his imagination free rein:

> I saw them together, Wordsworth, Haydon, Keats, in a room in Monkton's house on Queen Anne Street and imagined the sum of their every sensation and thought, and all the stuff, the feel of clothes, the creak of chairs and floorboards, the resonance in their chests of their own voices, the little heat of reputation, the fit of their toes in their shoes, and things in pockets, their separate assumptions of recent pasts and what they would be doing next, the growing tottering frame they carried of where they were in the story of their lives— all this as luminously self-evident as this clattering, roaring restaurant, and all *gone*. (170)

Joe here identifies with the embodied existence of Keats and his friends ('the resonance in their chests of their own voices') and acknowledges the significance of their interior lives; moreover, his description of the 'tottering frame ... of where they were in the story of their lives' resonates with Damasio's account of the endlessly renewed 'autobiographical self' which is created through the continuous updating of images of our sensations and perceptions. This passage thus opens a window onto an understanding of subjectivity which is materialist but non-reductionist, in line with Damasio's view that 'soul and spirit, with all their dignity and human scale' are simultaneously 'complex and unique states of an organism'.[21]

The novel also self-reflexively probes the implications of evolutionary psychology for fictional characterization and plot. Central to this exploration is the concept of self-deception, which was first mooted by Trivers and subsequently elaborated by Cosmides and Tooby. Trivers suggests that in the evolutionary past, the ability to deceive others would have been favoured by natural selection but so too would a capacity for self-deception, as the most effective way to deceive others is to deceive oneself. By drawing on experiments carried out for other purposes, evolutionary psychologists have found some support for the argument that self-deception could have an adaptive function.[22]

---

[21] Antonio Damasio, *Descartes' Error: Emotion, Reason and the Human Brain* (London: Vintage, 2006, p. 252.

[22] For example, Wright argues that the evidence from 'split brain' experiments supports the case for self-deception (pp. 274–5).

Moreover, they argue that it is not just endemic in wider social relations but is an intrinsic part of our most intimate relationships, where arguments make visible, in Trivers's words, 'whole landscapes of information [which] appear to lie already organized, waiting only for the lightning of anger to show themselves'.[23] This is McEwan's focus in the novel as he explores the part played by self-deception in the breakdown of the relationship between Joe and Clarissa. Joe is not only aware of the concept of self-deception but has written an article on it and draws on this expertise to analyse his own falling into this trap. The idea of self-persuasion first comes into his mind after an argument, as he wonders whether Clarissa has misinterpreted his relationship with Jed Parry to alleviate her own guilt about a lover. It is then a short step to a state of mind in which (self) knowledge floats in and out of consciousness, with Joe performing innocence before ransacking Clarissa's desk looking for evidence of her infidelity, rationalizing this on the grounds that 'I had a right to know what was distorting Clarissa's responses to Parry' (105). Joe's knowledge of evolutionary psychology allows him retrospectively to analyse his behaviour and to summarize the broader implications of the concept of self-deception, namely that 'we lived in a mist of half-shared, unreliable perception, and our sense data came warped by a prism of desire and belief which tilted our memories too'. He goes on to suggest that 'pitiless objectivity, especially about ourselves, was always a doomed social strategy', and his analysis suggests that it might also be a doomed aesthetic strategy, as it opens up a *mise-en-abyme* in which the reliability of both the narrator and the 'real' author of *Enduring Love* is called in question (180–1).

The idea that literary production might entail self-deception is explored in more detail in *Atonement*, a novel in which the literary concept of the unreliable narrator is given a biological twist. The theme is introduced in the opening scene in which we encounter thirteen year-old Briony Tallis trying to rehearse a play she has written. In his characterization of a young girl on the cusp of adolescence, McEwan knits together a number of themes drawn from evolutionary psychology. First, although Briony thinks of herself as a future writer she is still caught up in the 'pretend play' of childhood which according to

---

[23] Robert Trivers, *Social Evolution*, p. 420, quoted in Wright, p. 280.

Cosmides and Tooby is 'the original artistic medium' and the precursor of the fictional worlds we create as adults.[24] So we see Briony oscillating between sophisticated reflections on her artistic practice and fantasies of killing the cousin who has stolen her 'rightful role' in the play ('though she whimpered for mercy, the singing arc of a three-foot switch cut her down at the knees and sent her worthless torso flying').[25] Secondly, McEwan suggests that just at the point when she is beginning to recognize the limits of her ability to shape the world, Briony is also developing an awareness of the reality of other minds, as she asks herself:

> Was everyone else really as alive as she was? ... If the answer was yes, then the world, the social world, was unbearably complicated, with two billion voices, and everyone's thoughts striving in equal importance and everyone's claim on life as intense, and everyone thinking they were unique, when no one was. (36)

McEwan here dramatizes the moment when the theory of mind (TOM) proposed by evolutionary psychology becomes something of which we are consciously aware, adding a potentially 'unbearable' layer of complexity to our apprehension of the social world which surrounds us.[26] Finally, the concept of self-deception comes into play in the account of the 'crime' for which Briony feels she must atone, which is the false accusation that Robbie Turner has assaulted her cousin. The accusation stems from Briony's confused response to adult sexuality, as evidenced in the relationship between Robbie and her elder sister Cecilia. Witnessing scenes which she cannot understand, including their love-making in the library, she decides that Robbie is a sexual predator, 'a maniac treading through the night'. By constructing him as an all-purpose villain she is able to frame the complexities of adult behaviour in terms of a straightforward dichotomy between good and evil.

---

[24] L. Cosmides and J. Tooby, 'Does Beauty Build Adapted Minds? Toward an Evolutionary Theory of Aesthetics, Fiction and the Arts', *SubStance # 94/95*, 2001, pp. 24–5.

[25] Ian McEwan, *Atonement* (London: Vintage, 2001), p. 74. Further page references will be given within the main text.

[26] Theory of mind develops continuously from childhood to late adolescence and is closely linked with the language skills in which Briony excels. For an overview of theory of mind see H.M. Wellman, *The Child's Theory of Mind* (Cambridge, MA: MIT Press, 1992).

However, this entails a process of self-deception which the narrator anatomizes in detail, explaining that 'it was not simply her eyes that told her the truth...Her eyes confirmed the sum of all she knew and had recently experienced. The truth was in the symmetry, which was to say, it was founded in common sense' (169). Having intimated that Briony's perceptions are unreliable, McEwan jolts the reader further by revealing that she is not only the main protagonist but also the author of the novel we are reading. As in many postmodern novels, the effect of the switch between third and first person is to destabilize the relationship between fiction and reality but there is a crucial difference. Whereas postmodernism endlessly defers the question of reference, McEwan's novel is grounded in the belief that there is an empirical reality which is available, among other things, to scientific enquiry (represented here in Robbie's study of anatomy). However, human subjectivity is constitutively deceptive, for we have evolved to distort the evidence of our senses, impelled by a genetic unconscious which, according to Wright, is more radical than the Freudian unconscious because 'the sources of self-deception are more numerous, diverse, and deeply rooted' (324). Although there is no evidence that Briony has any knowledge of evolutionary psychology, her reflections touch on the implications of these theories for the writer, as she acknowledges the impossibility of achieving an 'impartial psychological realism' and recognizes that however much she strives to write a 'forensic memoir' it will be more or less 'self-serving' (372).[27]

The theories of evolutionary psychology also lie behind a shift in narrative mode as Briony distances herself from the modernist aesthetic which shaped her first account of her 'crime' in her novella *Two Figures by a Fountain*. After sending this to *Horizon* magazine she receives a letter from the critic Cyril Connolly which suggests that her work is too much influenced by Virginia Woolf, particularly in its focus on 'random impressions' and 'the vagaries and unpredictability of the private self' (312–13). These criticisms cohere with McEwan's views in his essay for *The Literary Animal*, where he argues that

---

[27] Briony's dilemma can be read as a refraction of a problem which is inherent in evolutionary psychology, which is that if unreliability is adaptive and is built into our thoughts and perceptions, how can we know that evolutionary psychology itself is anything more than a belief-formation with a selective advantage?

modernism is misguided in its excessive focus on subjectivity and its view of human nature as infinitely malleable, as in Woolf's famous claim that 'on or about December 1910, human character changed'.[28] The latter point is one which is picked up by Steven Pinker in *The Blank Slate* (2002) when he asserts that Woolf's comment, and modernism more broadly, represent 'a denial of human nature'.[29] For Pinker and other evolutionary psychologists, the concept of a universal, unchanging human nature is foundational, and while they acknowledge the existence of characteristics that arise in response to specific environments, their main interest is in the adaptive functions of human universals. As Wright puts it, 'evolutionary psychologists are trying to discern a second level of human nature, a deeper unity within the species' and to show 'how elegantly the theory of natural selection [...] reveals the contours of the human mind' (9, 11). McEwan makes similar points in *The Literary Animal* essay, arguing that 'behind the notion of a commonly held stock of emotion lies that of a universal human nature'. This concept has been 'reviled' for much of the twentieth century but as the climate of opinion is changing, it is time to refocus on the universal as well as the parochial dimensions of human nature, which are linked by McEwan to our genetic and cultural inheritance respectively (10).

These ideas are reflected in Part 2 of *Atonement*, as the novel turns away from the internal drama of Briony's consciousness, to explore that most intractable of human universals, war. This is a subject which has attracted considerable attention from evolutionary psychologists, most notably Tooby and Cosmides whose arguments are crisply summarized in Pinker's *How the Mind Works*.[30] Despite the multiple differences between modern warfare and Pleistocene conflict, they argue that

---

[28] Ian McEwan, 'Literature, Science and Human Nature', in Jonathan Gottschall and David Sloan Wilson (eds), *The Literary Animal: Evolution and the Nature of Narrative* (Evanston: Northwestern University Press, 2005, pp. 5–19. Further page references will be given within the main text.

[29] Steven Pinker, *The Blank Slate: The Modern Denial of Human Nature* (London: Penguin, 2002), p. 404. Further page references will be given within the main text. A shorter version of 'Literature, Science and Human Nature' was published in the *Guardian* in June 2001, so this may be a case of a writer influencing an evolutionary psychologist rather than the other way round. See Ian McEwan, 'The Great Odyssey', *Guardian*, 'Saturday Review' (9 June 2001), pp. 1–3).

[30] See Steven Pinker, *How the Mind Works*, (London: Penguin, 1997), pp. 514–17.

ancestral psychological mechanisms still 'interpenetrate modern war-fare'.[31] They contend that what they call 'coalitional aggression' was adaptive in the evolutionary past as it was a means of gaining access to women (the 'limiting factor' in male fitness) and they also highlight the crucial but largely unrecognized function of cooperation within these coalitions. The novel picks up on these themes as it charts Robbie's experiences in the retreat from Dunkirk. It highlights the (conscious or unconscious) importance to the combatants of their reproductive future, signalled in Robbie's focus on Cecilia as the reason 'why he must survive' and his response to her work on a maternity ward when he reflects that this was 'the unspoken call to [her] own future, the one she would share with him' (207). Later he reflects on the link between the personal costs of war and the collective investment in fatherhood which prompted the post-war baby boom: 'All around him men were walking silently with their thoughts, reforming their lives, making resolutions. If I ever get out of this lot... They could never be counted, the dreamed up children, mentally conceived on the walk into Dunkirk, and later made flesh.'[32] McEwan also probes the nature of the cooperative bonds which are forged in war when two corporals latch onto Robbie at the beginning of the march to Dunkirk because he has the skills in map-reading which they lack. Despite the fact that Robbie is frequently irritated by them he benefits from their practical support, especially when he begins to fall ill with septicaemia, at which point the fading out of his consciousness represents his assimilation to a 'coalition of aggression' driven by imperatives which exceed the survival of particular individuals.

In line with the shift in focus to human universals, Robbie is presented in this part of the novel as both individual and archetype, an Odysseus who fails to return (it is no coincidence that McEwan cites the *Odyssey* in *The Literary Animal* essay as the supreme expression of a human essence which transcends time). The narrative mode is one of sober realism and the style switches to short, declarative sentences

---

[31] John Tooby and Leda Cosmides, 'The Evolution of War and its Cognitive Foundations', *Institute for Evolutionary Studies Technical Report 88–1*, 1988, https://pdfs.semanticscholar.org/7f95/d9d117721df9e69b929b004d9d85ea6c560d.pdf [accessed 22 March 2018].

[32] McEwan's father, with whom he did not get on well, was part of the Dunkirk retreat and McEwan is perhaps making a personal atonement here, acknowledging his debt to the courage of his father and his father's generation.

which give the impression of relaying objective truths. We learn that Briony has, like McEwan, drawn on unpublished letters, journals, and reminiscences in the Imperial War Museum for her novel, a move which arguably strengthens its claim to represent universal rather than particular experience.[33] The contrast between the violence and physical suffering in this section of the novel and the focus on the private lives of the privileged in Part 1 also makes a broader point, suggesting that the inward turn of modernism constituted a refusal to engage with geopolitical realities and the impending war. This impression is reinforced by the juxtaposition of Briony's encounter with a dying soldier and her receipt of the letter from Connolly suggesting she should pay less attention to the private self. Realizing that the soldier is at the point of death, she plays along with his illusion that she is his fiancé, subduing her individuality to the archetypal role of the sweetheart, in a further acknowledgement of the force of universal human emotions.

McEwan's next novel, *Saturday*, is also concerned with human universals and with war as a defining human activity, this time in the context of 9/11 and the global 'war on terror'. The novel is set on 15 February 2003, the day of the march against the Iraq war, which took place against the backdrop of Tony Blair actively making the case for war. McEwan was fascinated by Blair, commenting in an unpublished note from 2005 'I'm not sure there's been a day in the last 10 years when I haven't thought about him—with fury, longing, amazement.' He also found him a tragic figure, 'an idealist in whose way so many obstacles—inner character—appeared' and compared his rise and fall to that of a Shakespearean hero.[34] In *Saturday*, he considers Blair's role from the point of view of Henry Perowne, the central consciousness in

---

[33] McEwan's efforts in this direction led to his being accused of plagiarism, despite the fact that he had fully acknowledged his use of published and unpublished material relating to the Second World War. The charge is ironic given that the novel is centrally concerned with re-examining and to an extent re-establishing the distinction between fiction and reality. As Natasha Alden points out, in this respect McEwan was challenging the blurring of the boundary between fiction and history which characterized the historiographic metafiction of the 1990s. See Natasha Alden, 'Words of War, War of Words: *Atonement* and the Question of Plagiarism', in Sebastian Groes (ed.) *Ian McEwan*, 2nd edition (London: Bloomsbury, 2013), pp. 57–69 (p. 59).

[34] Notes regarding Tony Blair (and plot notes for novel Saturday?), 8 May 2005, Box 29.3, Ian McEwan Archive, Harry Ransom Center, University of Texas at Austin.

the novel, a neurosurgeon who is also well-versed in neo-Darwinian theory and evolutionary psychology. Perowne is profoundly ambivalent about the prospect of war and watching Tony Blair on a bank of TV screens, he ponders the difficulty of knowing whether he is telling the truth about weapons of mass destruction. The problem is compounded by the fact that we cannot see him in the flesh, for as Perowne reflects, we have evolved to make decisions about trust on the basis of physical signs and 'tells' which are not available in the same way in an era of mass communication. Nonetheless, Perowne suspects that self-deception plays its part in Blair's advocacy of war, noting in this context that 'the first and best unconscious move of a dedicated liar is to persuade himself he's sincere. And once he's sincere, all deception vanishes.'[35]

As his notebooks show, in exploring the motivations of war McEwan draws directly on Pinker's discussion of violence in *The Blank Slate*. In Pinker's view, violence is an innate component of our genetically mediated human nature. Borrowing Dawkins's metaphor, he argues that as 'survival machines' we are programmed to neutralize any survival machine which poses a threat to our interests, even if this is another human, and that violence is 'a near-inevitable outcome of the dynamics of self-interested, rational social organisms' ((329). Stressing the continuities between conflict at the individual and group level, he argues that violence can also be the rational option at the level of the nation-state, citing a study which has shown that in most real-world conflicts 'the aggressor correctly calculated that successful invasion would be in its national interest' (319). Developing Hobbes's analysis of the dynamics of violence, he argues that competition, the need for self-defence, and honour are the main drivers of conflict, pointing out that fights over apparently trivial matters are critical to maintaining one's place in the male dominance hierarchy. He also argues that game-theoretical models can help us to understand the 'spiral of mistrust' which fuels the escalation from friction to war. In his notes on Blair, McEwan reads his actions in precisely these terms and underscores the fact that as a leader, he is strategizing with an eye to both national and personal honour: he writes that Blair 'gambled—played high stakes—to

---

[35] Ian McEwan, *Saturday* (London: Vintage, 2005), p. 329. Further page references will be given within the main text.

speak for invasion, and never to fall out in public with Bush— fantastically high game. If it had worked, he would have been a planetary demi-God.'[36]

Pinker's argument for the continuities between individual and group conflict becomes the structural principle of the novel, which explores the analogies between conflicts in Perowne's personal life and those on the wider geopolitical stage. By his own account, Perowne 'doesn't actually relish personal confrontation', but it is nonetheless built into his life through squash games with his closest colleague, Jay Strauss (85). In a detailed account of the game that takes place on this particular Saturday, it is emphasized that for the participants 'the irreducible urge to win, as biological as thirst' is mediated by fixed rules of engagement (113). However these are always open to interpretation, leading to disputes which are difficult to resolve in a sport where there is no referee, as happens on the morning in question, when a disputed call leads Strauss to ask Perowne if he is calling him a liar. This sets up an obvious parallel with the run-up to the Iraq war, when there were difficulties of interpretation regarding UN resolutions, suspicions of lies being told about WMDs, and no strong international 'referee' to mediate between the protagonists. From this point of view, the game can be seen as modelling the 'spiral of mistrust' which fuels war and as a comment on the relationship between the West and Saddam Hussein. However, it also alludes to the relationship between Tony Blair and George Bush, given that Strauss is not only an American but a 'man of untroubled certainties, impatient of talk of diplomacy', while his surname alludes to Leo Strauss, the philosopher who has been linked with the rise of neo-conservative politics who also argued that it was not always helpful for political leaders to be truthful.[37] Meanwhile, Perowne's mediation between Strauss and hospital bureaucracy reflects Blair's role in liaising between Bush and the United Nations. Reading across from the squash game to the Blair–Bush relationship, the novel draws attention to the fact that competition inheres *within* alliances as

---

[36] Notes regarding Tony Blair (and plot notes for novel Saturday?), 8 May 2005, Box 29.3, Ian McEwan Archive, Harry Ransom Center, University of Texas at Austin.
[37] For a discussion of the influence of Strauss's philosophy on neo-conservative thought see Shadia B. Drury, *Leo Strauss and the American Right* (Basingstoke: Palgrave Macmillan, 1999).

well as structuring 'coalitional aggression', and that this may further exacerbate conflict. The more fundamental implication, in line with Pinker's contention, is that the actions of political leaders are shaped by genetically mediated dispositions of which they may be unaware.[38]

As noted above, the squash game takes place in a context where the rules of competition, in sport and social life, are generally observed. However in an earlier encounter Perowne comes into contact with a section of society where the rules of engagement are different. He is involved in a minor accident which damages his expensive Mercedes and the BMW that belongs to a young man known as Baxter, whom he has previously seen coming out of a lap-dancing club and who appears to be a pimp and a drug-dealer. In the wake of the collision the two men square up to a fight but it soon becomes clear that the collision is not only between cars but between competing codes of behaviour. Baxter represents what McEwan terms 'the terrifying pride of the street', a phrase inspired by Pinker's account of the codes of honour which develop in those parts of society that are beyond the reach of the law.[39] As Perowne reflects in a passage taken almost word-for-word from *The Blank Slate*, 'drug dealers and pimps, among others who live beyond the law, are not inclined to dial nine-nine-nine for Leviathan' (in other words the police). The suggestion is that Baxter is in precisely this position and must defend his property by punching Perowne in the chest before ordering his sidekicks to slam him against a door. Perowne's response to this is telling: he eschews overt violence and opts

---

[38] McEwan's thinking about the alliance between Bush and Blair may also have been inspired by the discussion of political friendships in his friend Timothy Garton Ash's *Free World*. Garton Ash comments on the Bush–Blair summits in advance of the Iraq war and notes that 'friendship is the name diplomatically given to these relations between statesmen or stateswomen and, by two-way symbolic extension, to relations between the states they represent [...] These instant, speed-glued political "friendships" are interesting to observe. You wonder how genuine they can possibly be' (Timothy Garton Ash, *Free World: Why a Crisis of the West Reveals the Opportunity of Our Time* (London: Penguin, 2005), p. 7.

[39] McEwan's notes point to the influence of Pinker's thinking on the development of the novel. He reminds himself of 'Pinker's list of innate human shortcomings that must keep us from Paradise' before noting the 'Terrifying pride of the street. See Pinker in The Blank Slate. Without Leviathan.' Saturday (novel 2005) Research. Surgery observation notes with Dr Neil Kitchen, medical articles, Ray Dolan email printouts, notes, 2004, Container 14.3, Ian McEwan Archive, Harry Ransom Center, University of Texas at Austin.

instead to use his professional expertise to defeat his opponent. As a neurosurgeon, he recognizes that Baxter is in the early stages of Huntingdon's disease and confronts him with this knowledge, humiliating him in front of his associates and forcing his retreat in an act of symbolic violence. The dice are heavily loaded in this encounter, as Perowne possesses economic, social, and health capital whereas Baxter inhabits the semi-criminal underworld, has left school too early, and is suffering from a neurodegenerative disease thought to be caused by errors in genetic transcription. This imbalance sets up tensions which are never quite resolved in the novel. On the one hand, it could be argued that in identifying a dispossessed young man with an inherited disease, McEwan is making a progressive political point, underscoring the way in which genetic bad luck can ramify outwards and become associated with social and economic disadvantage. On the other hand, the way in which genetic disadvantage is layered over deprivation invites the charge that McEwan is reactivating the eugenic trope of the 'social problem group', that section of society which was thought to be congenitally incapable of rising above poverty and crime.[40] And if, as a number of critics have argued, Baxter represents Saddam Hussein, this too has extremely uncomfortable implications. By associating the Iraqi leader with a man suffering from a fatal genetic disorder the text invites the charge that it is reviving racist tropes of the degenerate other.[41]

These issues resurface at the end of the novel as Perowne looks out on the square outside his home, which with its Persian rugs and Bridget Riley prints embodies his position of economic, social, and cultural advantage. Insulated from the public space of the city, he thinks about

[40] For a discussion of the eugenicists' social problem group see Pauline M.H. Mazumdar, *Eugenics, Human Genetics and Human Failings: The Eugenics Society, its Sources and its Critics in Britain* (London: Routledge, 1991), chapter 5.

[41] In this context, it is notable that whereas Perowne-as-narrator promotes the traditional view of Huntingdon's as, in his words, 'biological determinism in its purest form', there was evidence at the time the novel was written which suggested that the age of onset and symptoms of the disease are modulated by the environment. See The US–Venezuela Collaborative Research Project, and Nancy S. Wexler. 'Venezuelan Kindreds Reveal That Genetic and Environmental Factors Modulate Huntington's Disease Age of Onset', *Proceedings of the National Academy of Sciences of the United States of America* 101.10 (2004): 3498–503. (PMC. Web. 29 May 2018, https://www.ncbi.nlm.nih.gov/pmc/articles/PMC373491/ [accessed 29 May 2018].

the successful office workers who populate the square by day ('men and women of various races . . . confident, cheerful, unoppressed') and contrasts them with the derelicts, 'the broken figures who haunt the benches' (272). From his perspective, the difference between those who succeed in the global marketplace and those who fail is a matter of genetic luck:

> It can't just be class or opportunities—the drunks and junkies come from all kinds of backgrounds, as do the office people. Some of the worst wrecks have been privately educated. Perowne, the professional reductionist, can't help thinking it's down to invisible folds and kinks of character, written in code, at the level of molecules. It's a dim fate, to be the sort of person who can't earn a living, or resist another drink, or remember today what he resolved to do yesterday. No amount of social justice will cure or disperse this enfeebled army haunting the public spaces of every town.   (272)

Perowne's gene-centric view of human nature echoes the bias of evolutionary psychology, which officially subscribes to the doctrine of gene-culture coevolution but in practice prioritizes genes as the determinants of human characteristics. His comment that 'no amount of social justice' will help the dispossessed also resonates with the opposition to progressive social intervention which is such a striking feature of evolutionary psychology. While this hostility may have originated in a desire to distinguish it from other branches of psychology, it develops in the hands of popularizers like Pinker and Ridley into a broad-brush critique of left-wing social policies. Pinker has been especially influential in this respect, arguing against 'social engineering' and linking the 'blank slate' model of human nature to the rise of totalitarian regimes in the USSR and communist China.[42] For Pinker, social arrangements must work with the grain of human nature, relying on 'time-tested' traditions associated with religion and family life to balance competing

---

[42] Hilary Rose points out that the evolutionary psychologists' concept of the Standard Social Science Model is a straw man, grounded in ignorance of the long history of thinking about the biological in the social sciences. In her view, evolutionary psychologists discredited the social sciences because they had ambitions to replace them, with the goal being social policies designed in line with their (retrogressive) understanding of human nature. Hilary Rose, 'Colonising the Social Sciences', in *Alas Poor Darwin*, pp. 106–28 (pp. 125–6).

interests. Such a non-interventionist approach can readily be aligned with neo-conservative politics and with neoliberalism, the dominant political rationality of the period when the novel was set. Indeed, as David Harvey has pointed out, the language of biological competition is frequently deployed to underwrite the principle of a competitive market economy, as advocates of neoliberal policies argue that 'in a Spencerian world [...] only the fittest can and do survive'.[43] Such a rhetorical move is evident in the conclusion to Matt Ridley's *The Origins of Virtue*, which extrapolates from the arguments of evolutionary psychology to make the case for free market economics. Although McEwan's novel does not pursue these ideas in detail, it is distinctly ambivalent about the political entailments of evolutionary psychology. For example, Perowne's scepticism about the effects of social justice is immediately undercut by his acknowledgement that he is 'no social theorist': his views are relativized as those of a blinkered materialist who knows little about the social sciences. On the other hand, his pessimistic view of Islamic terrorism and the impending war seems to be endorsed by the text, reflecting the 'tragic vision' of human nature espoused by Pinker, who at this point in his career sees humanity as always 'teetering on the brink of barbarism' (289).[44]

The novel also considers the role of the arts in politically charged times and in this respect is in close conversation with Matthew Arnold, whose poetry and criticism were also written against a backdrop of social and political unrest.[45] The issue comes to the fore in the climactic

[43] David Harvey, 'Neoliberalism as Creative Destruction', *Annals of the American Academy of Political and Social Science*, 610 (2007), pp. 21–44 (p. 34). Spencer is invoked as the originator of the phrase 'survival of the fittest', his term for Darwin's concept of natural selection. Spencer's views are also pertinent as he believed that organisms have a tendency to self-preservation which is expressed as rational self-interest: in this respect he can be seen as a precursor of evolutionary psychology.

[44] Perowne's views also reflect McEwan's public statements about religious terrorism. In one interview, for example, he argues that 'what we need more in the world is doubt; more scepticism, less crazed certainty. I feel that religious zeal, political zeal, is a highly destructive force. People who know that answer and are going to impose it on everyone else, I think, are terrifying people' ('faith and doubt at ground zero', https://www.pbs.org/wgbh/pages/frontline/shows/faith/interviews/mcewan.html [accessed 3 January 2018].

[45] Arnold was disturbed by the 1866 Hyde Park riots, which broke out when the Reform League was prevented from assembling in the park to express criticism of legislation which would extend the franchise but still deny the vote to working people. He alludes to the riots In *Culture and Anarchy*, expressing his fear of social breakdown

scene of the novel when Baxter forces his way into Perowne's home, seeking revenge for his earlier humiliation. He threatens Perowne's wife with a knife, demands that his daughter Daisy takes off her clothes, then seeing a copy of her first book of poems, demands that she recites one of them. At a hint from her grandfather she recites Matthew Arnold's 'Dover Beach' instead, calming and disarming Baxter. The poem functions in part as an ironic commentary on human progress, a reminder of the perennial nature of war and conflict, whether motivated by loss of faith, as Arnold's poem suggests, or by faith itself, as in the context of Islamic terrorism. However, Baxter's reaction to the poem also illuminates a view of the purposes of art which differs significantly from that of Arnold. Famously, Arnold defined culture in terms of 'sweetness and light'.[46] For him, art was a crucible for human perfection and was inseparable from social progress, an idea he developed by arguing that 'the State' is the 'organ of our collective best self, of our national right reason' (97). Reason, however, is not the most salient factor in Baxter's change of heart. Some critics have argued that this mood change is implausible, but it becomes intelligible when viewed in the light of the function of art as theorized by evolutionary psychologists.[47] Cosmides and Tooby are among those who have argued that art is an evolved adaptation: they have suggested that fiction, specifically, enhances our capacity for counterfactual reasoning.[48] Taking a different view, Ellen Dissanayake describes art as a form of 'making special', originating in the 'protoconversations' between infants and mothers which attune infants to social priorities.[49] Building on her

---

and arguing that the development of a sense of shared authority, embodied in the state, was the best means of averting this.

[46] Matthew Arnold, *Culture and Anarchy*, ed. J. Dover Wilson (Cambridge: Cambridge University Press, 1971), p. 54. Further page references will be given within the main text. See Helen Small, *The Value of the Humanities*, pp. 81–8, for a detailed discussion of the significance of Arnold's phrase.

[47] For example, Greg Garrard argues that there is something almost parodic about Baxter's transformation into 'the ideal subject of liberal humanist literary theory'. See Garrard, 'Ian McEwan's Next Novel and the Future of Ecocriticism', *Contemporary Literature* 50 (2009), 695–720 (715).

[48] Cosmides and Tooby, 'Does Beauty Build Adapted Minds', p. 20.

[49] Ellen Dissanayake, *What is Art For* (Seattle: University of Washington Press, 1988), p. 64; *Art and Intimacy: How the Arts Began* (Seattle: University of Washington Press,

work, the literary Darwinist Brian Boyd argues that because humans have such a long period of childhood dependency, the ability to share and shape the attention of others offers a significant survival advantage. For him, art has an individual function, enhancing capacity for sharing and commanding attention, and a social function, increasing 'social attunement' and social cohesion.[50] The impact of the poem on Baxter can be seen as an illustration of these theories. Its rhythms command his attention, as though reactivating the 'protoconversations' of childhood, a notion supported by his comment that the poem makes him 'think about where he grew up', while Perowne notes that in reading it Daisy employs 'the seductive, varied tone of a storyteller entrancing a child' (221–2). Looking back, Perowne concludes that the poem worked as a kind of spell or trick, prompting an experience of 'shared attention' which defused social tension (278).

In McEwan's subsequent fiction there is a drift away from the interpretive framework of evolutionary psychology, which is alluded to but which no longer has a primary role in the shaping of character and plot. For example in *Solar*, the disorder in a boot room is interpreted in terms of rational self-interest (as it is in McEwan's blog post on this topic) but the novel swerves away from any in-depth reflection on the link between self-interest as an evolutionary strategy and the anthropogenic destruction of the planet.[51] McEwan's subsequent novels, *Sweet Tooth* and *The Children Act*, move in new directions, as *Sweet Tooth* self-referentially explores McEwan's early career as a writer while *The Children Act* examines the relationship between science, law, and religious faith. It seemed that McEwan had lost interest in biological explanations of human behaviour but a more recent novel suggests that it was not so much a loss of interest as a loss of faith in neo-Darwinian orthodoxies. *Nutshell* breaks decisively with neo-Darwinian narratives and opens up a radically different perspective on human nature and development. McEwan has described the process

---

2000), p. 7, quoted in Brian Boyd, 'Evolutionary Theories of Art' in *The Literary Animal*, pp. 147–76.

[50] Brian Boyd, 'Evolutionary Theories of Art', p. 153.

[51] See Ian McEwan, 'A Boot Room in the Frozen North', http://www.capefarewell.com/explore/215-a-boot-room-in-the-frozen-north.html [accessed 25 March 2018].

of writing the novel in terms of 'abandoning' the laws of biology and has located its origin in a moment of intuition when he was chatting to his pregnant daughter-in-law: '"We were talking about the baby and I was very much aware of the baby as a presence in the room," he recalls.'[52] Yet McEwan's intuition does not so much lead him away from biology as open up a landscape which resonates with the scientific and philosophical insights deriving from recent research in epigenetics. In this respect, *Nutshell* can productively be read alongside the philosopher Catherine Malabou's recent work on epigenesis and epigenetics. Malabou's work constitutes one of the most sophisticated recent attempts to think through the relationship between the biological and the transcendental. In so doing, she opens up a novel perspective on the 'hard question' of consciousness earlier raised in *Enduring Love*.

In her most recent book, *Before Tomorrow: Epigenesis and Rationality*, Malabou goes back to Kant's *Critique of Pure Reason* and reads it in relation to eighteenth-century debates in which epigenesis (the theory of development through the addition of parts which are born from each other) is opposed to preformationism (the belief that the embryo is a fully constituted being whose growth entails unveiling already-formed parts). In Malabou's reading of Kant, transcendental reason is itself understood in terms of epigenesis, as folded into the temporality of epigenetic becoming, the biological and the logical being thereby co-articulated. As she suggests, there is a clear parallel between the eighteenth-century debates over epigenesis versus preformationism and twenty-first-century controversies over the relative importance of genetic and epigenetic factors in development. At stake in both contexts is the articulation of the biological and the logical and the question of how far the developing organism is malleable and responsive to environmental cues; also the extent to which development can be understood in terms of autonomous self-formation. For Malabou, the epigenetic perspective enables an intrication of the biological and the logical and an understanding of human nature as formed within a network of relationships extending from the single cell to the organism to the wider environment. Moreover, she suggests that at the intersection of 'transcendental epigenesis' and biological self-organization, we can locate 'the point of emergence of auto-affected, interpreting subjectivity'.[53] The subject is formed 'through the act of receiving its own spontaneity', as transcendental epigenesis inaugurates what Malabou calls 'the *adventure of subjectivity*' (93–4, her italics).

McEwan's novel opens up to this adventure. As a re-working of *Hamlet* it is by definition engaged with ontological issues ('to be or not to be') but these are given a distinctive twist by being displaced on the foetus-narrator. Unlike Hamlet, he is a' fresh-faced empiricist' who takes it as self-evident that thought is intricated with biology and that 'felt sensations are the beginning of the invention of the self'.[54] His malleability is suggested in the conceit of the sophisticated palate he has acquired during months of imbibing wine through his mother's placenta, a talent which is a clear (if unfortunate) example of an epigenetically-mediated disposition.[55] In addition, the novel offers an intricate and subtle reading of the formation of the self, one which qualifies the narrator's opening claim that he is a blank slate by sketching a developmental landscape in which what Malabou describes as 'a formative force' (represented in the novel as 'knowledge we simply arrive with') grounds biological self-organization and the emergence of the productive subject. The auto-affective dimension of subjectivity is represented in the image of a silkworm spinning its own cocoon: 'many weeks ago, my neural groove closed upon itself to become my spine and my many million young neurons, busy as silkworms, spun and wove from their trailing axons the gorgeous golden fabric of my first idea' and the self-forming aspect of subjectivity is further underscored as the narrator describes himself as 'a slate that writes upon itself as it grows

[53] Catherine Malabou, *Before Tomorrow: Epigenesis and Rationality*, trans. Carolyn Shread (Cambridge: Polity Press, 2016), p. 90. Further page references will be given within the main text.

[54] Ian McEwan, *Nutshell* (London: Jonathan Cape, 2016), p. 46. Further page references will be given within the main text.

[55] In this respect, epigenetics resonates with the traditional concept of maternal impressions, whereby it was thought that a pregnant women's emotions could have a physical effect on her child. In the novel, the narrator is strongly affected by his mother's moods as when he responds in this way to her anger: 'Trudy's anger is oceanic—vast and deep, it's her medium, her selfhood. I know it in her altered blood as it washes through me, in the granular discomfort where cells are bothered and compressed, the platelets cracked and chipped. My heart is struggling with my mother's angry blood' (p. 77).

by the day', an image which recalls Malabou's account of transcendental epigenesis in terms of self-interpretation (2).

Nutshell thus inserts itself in the space between hard-wired genetic determinism and the blank slate model of human nature which evolutionary psychologists viewed with such suspicion, and in so doing it opens up the implications of epigenesis and epigenetics for our understanding of human agency in development and beyond. As we have seen, epigenetics construes the organism as dynamic and responsive to changing contexts, a perspective which, as Malabou has argued in her previous work, implies a considerable degree of agency and autonomy.[56] Her view is echoed from a different philosophical perspective by John Dupré, who links epigenetics with niche construction and cultural evolution to make the case for a human agency which exceeds genetic predetermination.[57] McEwan's novel resonates with these perspectives, although there is no direct evidence that he is writing in response to revisionist twenty-first-century biology. Opening up a window on the womb, Nutshell depicts an unborn child already complexly embedded in its environment, accessing food, drink, and emotions through his interactions with the maternal body, while responding as well to the barrage of information which comes to him via the BBC, an important mediator of the wider world. In this context he develops a proto-subjectivity capable of both shaping and acting on an intention. As he puts it, capturing the recursive nature of his decision-making: 'After all my turns and revisions, misinterpretations, lapses of insight, attempts at self-annihilation, and sorrow in passivity, I've come to a decision. Enough' (192). Unlike Hamlet, he acts decisively, initiating an early labour which prevents Trudy and Claude escaping from the police. His birth avenges his father's death, in a dramatic example of foetal agency.

Perhaps the most striking aspect of Nutshell is the narrator's repudiation of explanations of human behaviour in terms of genetic self-interest. As we have seen, these explanations are at the core of evolutionary biology but in a Guardian podcast on Nutshell McEwan

---

[56] See Catherine Malabou, What Shall We Do With Our Brain?, trans. Sebastian Rand (New York: Fordham University Press, 2008), p. 75.

[57] John Dupré, Processes of Life: Essays in the Philosophy of Biology (Oxford: Oxford University Press, 2012), p. 289.

describes them as 'interesting just so stories', here alluding to Stephen Jay Gould's critique of this aspect of evolutionary psychology.[58] Gould argued that when evolutionary biologists try to explain form and behaviour by reconstructing history, they are telling stories as fictive as Kipling's *Just So Stories*, and he points out that such speculations have clear socio-political implications, for example in relation to gender roles.[59] A related point is made by the anthropologist Susan McKinnon, who argues that the theories of evolutionary psychology have gained cultural traction because they 'gather into one grand narrative a number of beliefs that are central to Euro-American culture', including the innateness of gender difference, the naturalness of neoliberal economic values, and the determinant force of genes.[60] The 'just so' story that *Nutshell* picks up on is one which is of obvious relevance to the narrator, which is that that the genetic interests of parents and children may be fundamentally at odds. As Wright explains, the reasoning behind this argument is that while a child is 100 per cent invested in her own genes, parents have a 50 per cent investment in the genes of all their children, including prospective ones (166). From this point of view, it is in the interest of a mother to ration the nurture she gives to any one child, or as McEwan's narrator puts it, 'pregnant women must fight the tenants of their wombs. Nature, a mother herself, ordains a struggle for resources that may be needed to nurture my future sibling rivals.' On the other hand, from an evolutionary point of view, it would be in the narrator's father's interest to trick another man into raising his child while he distributes his 'likeness among other women' (33). However, having rehearsed these propositions, the narrator then turns against them, rejecting them not on the grounds of lack of evidence, but because they are 'so bleak, so loveless... Too much to bear, too grim to be true. Why would the world configure itself so harshly? (34) This response accords with Marilynne Robinson's assessment of the inadequacy of a worldview which rests on 'a definition of

---

[58] 'Ian McEwan on his novel *Nutshell*', Guardian Books Podcast, 2 September 2016, https://www.theguardian.com/books/audio/2016/sep/02/ian-mcewan-on-his-novel-nutshell-books-podcast [accessed 25 March 2018].
[59] Stephen Jay Gould, 'Sociobiology: the art of storytelling', *New Scientist*, 16 November 1978, pp. 530–3.
[60] Susan McKinnon, *Neo-liberal Genetics: the Myths and Moral Tales of Evolutionary Psychology* (Chicago: Prickly Paradigm Press, 2005), p. 144.

the mind, therefore the human person, which tends to lower us all in our own estimation'.[61] It also invites us to reflect on the broader impact of McEwan's engagement with evolutionary psychology on his fiction. In general, critics have been respectful of this strand in McEwan's work, with Dominic Head, for example, seeing his interest in evolutionary psychology as seamlessly integrated with the 'quests for selfhood' that underpin his fiction.[62] However, there is a case to be made that the reductionism and eliminationism of evolutionary psychology pose particular difficulties for a writer of fiction and that this science has, accordingly, had a depressant effect on McEwan's writing. Robinson's critique of evolutionary psychology, written from the point of view of a theist but also that of a Pulitzer prize-winning novelist, is illuminating in this respect. In an essay in *The Givenness of Things* she challenges evolutionary psychology's reductionist assumption that there is 'an essential and startlingly simple mechanism behind the world's variety'. She also contests its eliminationism, that is, its focus solely on behaviours which can be linked to survival and reproductive fitness. Taken together, these premises suggest that 'there are tiers to existence or degrees of it, as if some things, though manifest, are less real than others and must be excluded': by this means 'the main work of human consciousness for as long as the mind has left a record of itself' is devalued.[63] Strikingly, she then turns to the significance of epigenetics for our self-understanding, elegantly summarizing the challenge it offers to 'Darwinian cost–benefit analysis':

> Now we know that chromosomes are modified cell by cell, and that inheritance is a mosaic of differentiation within the body, distinct in each individual. Therefore the notion that one genetic formula, one script, is elaborated in the being of any creature must be put aside, with all the determinist assumptions it has seemed to authorise.
>
> (14–15)

Unlike evolutionary psychology, epigenetics offers an understanding of the human in terms of individuation, legitimating our intuitive 'sense of

---

[61] Marilynne Robinson, *Absence of Mind*, p. 32.

[62] Dominic Head, *Ian McEwan* (Manchester: Manchester University Press, 2013), p. 19.

[63] Marilynne Robinson, *The Givenness of Things: Essays* (New York: Picador, 2016), pp. 11–12. Further page references will be given within the main text.

selfhood' (15). This is a biological perspective which is congenial with Robinson's novelistic commitment to exploring 'the negotiations the mind makes with itself' (91).

Pursuing this line of thought, it could be argued that evolutionary psychology's abstract models of human behaviour have had too great an influence on McEwan's fiction and in particular, that they have tilted the balance of his characterization too far towards the universal, even the generic. Comments from James Wood and John Banville on the schematic nature of the characterization in *Enduring Love* and *Saturday* would tend to support this view.[64] However, in a recent discussion of short stories which inspire him, McEwan has expressed a renewed concern with subjectivity, arguing that they illustrate:

> fiction's generous knack of annotating the microscopic lattice-work of consciousness, the small print of subjectivity. Both are third-person accounts that contain a pearl of first-person experience [...] In the act of recognition, the tight boundaries of selfhood give way a little. This doesn't happen when you learn what a Higgs boson does.[65]

Here he suggests that it is the alchemical fusion of first- and third-person perspectives which enables that identification with others which, in his view, is literature's principal contribution to ethical life. Such an understanding is entirely compatible with that of the modernist writers from whom McEwan distanced himself in *Atonement* and the *Literary Animal* essay. Indeed, the account of fiction as annotating 'the microscopic lattice-work of consciousness' resonates with Woolf's description of the novelist recording 'the shower of innumerable atoms' that fall upon the mind: in both cases a scientific metaphor is used to

---

[64] Wood suggests that McEwan's narrative designs 'do not open up but close off': see 'Containment: Trauma and Manipulation in Ian McEwan' in *The Fun Stuff and Other Essays* (London: Vintage, 2014), p. 185. Banville is more blunt, arguing 'that it happens occasionally that a novelist will lose his sense of artistic proportion, especially when he has done a great deal of research and preparation. I have read all those books, he thinks, I have made all these notes, so how can I possibly go wrong?' See John Banville, 'A Day in the Life', *New York Review of Books* 26 May 2005, http://www.nybooks.com/articles/2005/05/26/a-day-in-the-life/[accessed 25 March 2018].

[65] 'Ian McEwan: when faith in fiction falters—and how it is restored', *Guardian* 16 February 2013, https://www.theguardian.com/books/2013/feb/16/ian-mcewan-faith-fiction-falters [accessed 10 February 2018].

reawaken, rather than to close down, a sense of the complexity of the mind.[66] A renewed commitment to subjectivity is also signalled by the fact that In *Nutshell*, the narrator's emerging consciousness is aligned with that of Hamlet, whose characterization McEwan describes as the most 'extraordinary depiction of a consciousness ... an achievement in rendering thought processes perhaps never bettered since'.[67] Despite his confinement, the narrator's mind opens onto the 'infinite space' invoked in the novel's epigraph, an image of interiority which exceeds the interpretive rubric of evolutionary psychology. In this relatively short novel, McEwan reaffirms the irreducible value of fiction's exploration of the interplay between third- and first-person modalities of experience.

[66] Virginia Woolf, 'Modern Fiction', 1921, http://gutenberg.net.au/ebooks03/0300031h.html#C12 [accessed 25 March 2018].

[67] Talk on Consciousness, Wellcome Trust, 2 Ocober 2008, Container 28.4, Ian McEwan Archive, Harry Ransom Center, University of Texas at Austin.

| 4 |

Clone Lives

Eva Hoffman and Kazuo Ishiguro

T he birth of the cloned sheep Dolly took the scientific world by
surprise, as cloning was no longer in the mainstream of molecular
biology in the 1990s.[1] The field had been hit by repeated accusations of
fraud and had retreated from prestigious molecular laboratories to the
obscurity of research in agricultural facilities such as the Roslin Institute in Scotland. Nonetheless when Dolly's birth was announced the
response was overwhelming, transforming the lives of the team led by
Ian Wilmut and prompting fevered speculation about the possibility of
human cloning. This was not Wilmut's goal (his research aimed at
creating tissues for medical research) but Dolly's existence showed that
there was in principle no reason why it could not be done. As Wilmut
argued in his book *The Second Creation*, Dolly broke through the so-
called natural boundaries that were assumed to constrain human
action, to the extent that nothing seemed biologically impossible.[2]

---

[1] For an excellent and thorough account of the history of cloning, see Gina Kolata,
Clone: The Road to Dolly and the Path Ahead (London: Penguin, 1997). Ishiguro read and
made notes on this as part of the background research for *Never Let Me Go*: see Box 5.4
Ishiguro Never Let Me Go: 'Ideas as they Come', Notebook 1, Jan '01–Sept '02, sheet dated
2001, Kazuo Ishiguro Archive, Harry Ransom Center, University of Texas at Austin.
[2] Ian Wilmut, Keith Campbell and Colin Tudge, *The Second Creation: The Age of
Biological Control by the Scientists Who Cloned Dolly* (London: Headline Book Publishing,

Genetics and the Literary Imagination. Clare Hanson, Oxford University Press (2020). © Clare Hanson.
DOI: 10.1093/oso/9780198813286.001.0001

Her creation confirmed new facts about biological indeterminacy and in this respect contributed to the epistemic shift which took place around the millennium, as gene-centric neo-Darwinism was displaced by an understanding of living systems as plastic, autopoietic, and indeterminate. It also had significant social implications, breaking the link between heterosexuality and reproduction and thereby opening up the prospect of a reconfiguration of familial norms. This chapter explores the way in which literary fiction engages with these possibilities and with the biopolitical imaginary of the twenty-first century, focusing on Eva Hoffman's overlooked novel *The Secret* (2000) and (at the other end of the scale in terms of its profile) Kazuo Ishiguro's *Never Let Me Go*.[3]

Dolly's genealogy is complicated. The mammary cells used to create her were from a cell line derived from the udder of an elderly Finn Dorset ewe, the cell line having been cryopreserved by PPL Therapeutics, a partner in Wilmut's research.[4] By activating the cultured cells at a specific stage in the cell cycle (the 'resting stage'), the team were able to fuse them with donor eggs from a Scottish Blackface sheep and the resulting embryos were subsequently transferred into surrogates. Because the eggs had their nucleus removed, the transferred mammary cell provided 100 per cent of the nuclear DNA for the resulting offspring, with the result that Dolly was immediately recognizable as a Finn Dorset rather than a Scottish Blackface ewe. Dolly thus had three female parents and sexual reproduction was sidelined in her making. As Wilmut writes, 'her birth was other-worldly—iterally a virgin birth; or at least one that did not result directly from an act of sex' (233). Her

---

2000), p. 17. Further references will be given within the main text. The geneticist Lee Silver responded to Dolly's birth in a similar vein, commenting that 'It basically means that there are no limits. It means that all of science fiction is true. They said it could never be done and now here it is, done before the year 2000' (quoted in Kolata, p. 32).

[3] *Never Let Me Go* was a best-selling novel and a successful film version was released in 2010, with a screenplay by Alex Garland. Ishiguro was awarded the Nobel Prize in Literature in 2017, further raising his profile. For his Nobel acceptance speech see https://www.nobelprize.org/prizes/literature/2017/ishiguro/lecture/ [accessed 12 April 2019].

[4] For an account of the funding of the research that led to Dolly, see Wilmut et al., p. 8. As Sarah Franklin argues, Dolly was a manifestation of old and new forms of capital: capital as stock for breeding and the 'promissory capital' of bio-futures markets. See Sarah Franklin, *Dolly Mixtures: The Remaking of Genealogy* (Durham and London: Duke University Press, 2007), p. 47.

creation not only dislodged the biological categories that underpinned the reproductive order but also demonstrated for the first time that cellular differentiation could be reversed. As we have seen in previous chapters, prior to the advent of Dolly it was thought that as cells differentiate and become specialized (as they become skin cells, liver cells, and so on), they lose the ability to become other types of tissue. Cell differentiation was thought to be unidirectional and irreversible, just as, according to Francis Crick's Central Dogma of molecular genetics, sequencing information could not be transferred back from protein to DNA. However, Wilmut's team showed that specialized adult cells (like the mammary cell used for Dolly) could be reprogrammed or 'dedifferentiated' and that the ability to become another type of cell was always there in latent form. As Catherine Malabou points out, the techniques of cloning take us back to potentials that were present in primitive organisms but which were thought to have disappeared. As she writes, 'asexual reproduction and regeneration [...] represent ancient forms of life realized by the state of the art technologies of therapeutic and reproductive cloning'.[5]

Sarah Franklin has suggested that the scientific scandals around cloning intensified the understanding of the clone as being both copy and fake. The clone is generally viewed as inferior and subordinate because it lacks separation from the original and is, accordingly, thought to be without an 'individually defining substance'.[6] In its identity with the original, it is also clearly a variant on the uncanny double which represents aspects of the self which have been forgotten or repressed. Hoffman's novel *The Secret* exploits these associations to register the affective experience of second-generation Holocaust survivors, an issue which is also discussed at length in her non-fiction,

---

[5] Catherine Malabou, 'One Life Only: Biological Resistance, Political Resistance', trans. Carolyn Shread, *Critical Inquiry* 42 (Spring 2016), 429–38, available at https://criticalinquiry.uchicago.edu/one_life_only/ [accessed 12 April 2019].

[6] Franklin, *Dolly Mixtures*, pp. 26–7. The cloning scandals of the 1970s are detailed in Kolata, chapters 5 and 6. The first involved a fictional account of human cloning by the journalist David Rorvik, which was subsequently revealed to be an elaborate hoax. The second involved a gifted scientist, Karl Illmensee, who claimed in the summer of 1979 to have cloned three mice but was subsequently charged with fraud. Although he persisted in trying to clear his name his career never recovered.

notably in *Lost in Translation* and *After Such Knowledge*.[7] In the latter, she argues that the second generation lacks existential primacy because 'the incontrovertible significance of survivors' histories' instilled in their children the feeling that their experiences could never measure up to 'the size and import of their parents' ordeals'.[8] In addition, she suggests that the second generation has grown up with the uncanny in the sense that they are haunted by memories which are not their own, inheriting shadows which are both alien and deeply familiar. In *The Secret* the second generation's feelings of inauthenticity are translated into the terms of genetic replication, so that when the narrator, Iris, discovers she is a clone she feels that her sense of herself as 'a young girl with her very own, unique self' is an illusion and that she is 'nothing more than a Xerox of her cellular matter, an offprint of her genetic code'.[9] In addition, she is haunted by the feeling that her experiences are never exactly hers. In an allusion to the transpositions which often characterize post-Holocaust family ties, she suggests that 'in me, in my body, time repeated itself in some ghastly glitch [...] I looked at [my mother's] face, which was going to become mine, and knew that I had already been' (65).

After the success of Ira Levin's thriller *The Boys from Brazil*, in which the Nazi doctor Josef Mengele creates clones of Adolf Hitler, clones became emblematic of the dream of biopolitical control which animated Nazi genocide.[10] However, by focusing on the clone-as-victim rather than on the perpetrators, Hoffman's novel directs attention to the fact that the Holocaust entailed a kind of genealogical rupture for

[7] Hoffman's work has also been important to Marianne Hirsch's theorization of postmemory, a term she coined to capture the relationship of second-generation Holocaust survivors to the personal and collective trauma of their parents. For Hirsch's reflections on *After Such Knowledge* see *The Generation of Postmemory: Writing and Visual Culture After the Holocaust* (New York: Colombia University Press, 2012). For a discussion of postmemory in *The Secret* see Elizabeth Kella, 'Matrophobia and Uncanny Kinship: Eva Hoffman's *The Secret*', *Humanities*, 79 (2018), doi:10.3390/h7040122, https://www.mdpi.com/2076-0787/7/4/122/htm [accessed 13/04/2019].

[8] Eva Hoffman, *After Such Knowledge: A Meditation on the Aftermath of the Holocaust* (2004) (London: Vintage, 2005), p. 69. Further references will be given within the main text

[9] Eva Hoffman, *The Secret* (2001) (London: Vintage, 2003), p. 61. Further references will be given within the main text.

[10] See Ira Levin, *The Boys from Brazil* (1976) (London: Corsair, 2011). The film was released in 1978.

the second generation. As Foucault emphasizes in his analysis of Nazi biopolitics, the murder of the Jews was represented in terms of the elimination of a biopolitical threat to the population from a 'degenerate' race.[11] Or as Hoffman puts it in *After Such Knowledge*, the victims of the Holocaust were assaulted 'not for reasons of state, or as enemy combatants, but simply because of who they were' (43–4). One of the consequences of this deep racism, this attack on the inmost being of the victims, was that the relationship between the survivors and their children became impossibly freighted. As Hoffman writes, the parents 'invested so much in these children, and imbued them with so much yearning. To replace—revive—the dead ones; to undo the losses; to repair the humiliations wrought by the abusers' (63). Children like Hoffman (who was given the names of her murdered grandmothers) became 'votaries on the altar of the Shoah, their own lives dedicated to their hurt parents and the perished' (64).[12] In a reversal of the sequence of generation they are called to incarnate 'something [...] that we thought' had been dead' (65). In *The Secret*, Iris's cloning works as a metaphor for this psychic break, in which the second generation are detached from the chain of generation because they are bound to processes of reparation: accordingly, in relation to her biological family, Iris sees herself as 'a revenant, a spectre made flesh' (132).

The novel also addresses the cultural anxieties which swirled around cloning at the turn of the millennium and which were articulated by Jean Baudrillard in a series of influential essays. Baudrillard's frequently hyperbolic rhetoric reflects the degree of concern the clone provoked as an image of a new kind of human and as an emblem of broader social change. Drawing on August Weissman's evolutionary theory and on Freud's 'Beyond the Pleasure Principle', Baudrillard argues that cloning is a sign of biological and cultural regression, a form of species suicide

---

[11] See Michel Foucault, (2003) *Society Must Be Defended: Lectures at the Collège de France, 1975–76*, trans. David Macey (London: Penguin, 2004), p. 255. See also Paul Weindling, *Health, Race and German Politics* (Cambridge: Cambridge University Press, 1993) for a detailed analysis of the category of the degenerate in Nazi Germany.

[12] Dina Wardi describes her work with such second-generation survivors in *Memorial Candles: Children of the Holocaust* (The International Library of Group Psychotherapy and Group Process) (London: Routledge, 1992).

which he dubs, provocatively, 'the Final Solution'.[13] For him, cloning entails a desire for immortality which is bound to the death drive because it signals a longing for sameness and in-differentiation; this, in turn, threatens to reverse the evolutionary process. For Baudrillard as for Weissman, the advent of sexual reproduction is crucial to the development of higher organisms and in abandoning it we risk biological 'involution': 'where once living creatures strove, over millions of years, to pull themselves free of [...] incest and primitive entropy, we are now, through scientific advances themselves, in the process of creating precisely these conditions. We are actively working at the "disinformation" of our species through the nullification of differences' (8). In this view, sexual reproduction is aligned with variation, adaptability, and evolutionary progress, whereas cloning is linked with the primitive, the perverse, and the pathological. Paradoxically, it is 'sexual liberation' which has undermined sexual difference, as 'first, sex was liberated from reproduction; today it is reproduction that is liberated from sex through asexual, biotechnological modes of reproduction' (Final Solution, 10). Sexual liberation thus marks the end of the 'real' sexual revolution (the advent of sexuality in the world of living things) and sex becomes 'a useless function'. This forceful articulation of the threat cloning poses suggests an underlying hostility to the social changes that were associated with late-twentieth century feminist, lesbian, and gay politics: for Baudrillard, cloning is a metaphor for a breaching of the constraints of heterosexual family forms, as new modes of reproduction promise to facilitate new forms of parenting and kinship. On a related issue, he argues that cloning will dislodge the Oedipal structures that underpin normative subjectivity. As genealogy is re-spatialized and re-temporalized, new modes of attachment will generate a 'whole range of potential problems'; Oedipal psychology will be turned upside down when conflicts 'no longer center on the child and his or her parents but on the original and its double' (27).

---

[13] Jean Baudrillard, *The Vital Illusion* (The Wellek Library Lectures), ed.by Julia Witwer (New York: Columbia University Press, 2001), p. 9. Further references will be given in the main text. See also August Weismann, *Essays upon Heredity* (Oxford: Oxford University Press, 1889), esp.org, http://www.esp.org/books/weismann/essays/facsimile/contents/weismann-essays-1-a-fm.pdf [accessed 13 April 2019]; Sigmund Freud, *The Standard Edition of the Complete Psychological Works of Sigmund Freud*, vol. 18, ed. James Strachey (London: Hogarth Press and the Institute of Psychoanalysis, 1953–74).

The Secret picks up on these fears but suggests that they are misplaced. In relation to the putative link between cloning and the death drive, Hoffman suggests that what is at stake is not species suicide but the way in which cloning, like other reproductive technologies, forces a confrontation with human origins. When Freud theorizes the death drive and represents the living being as emerging from and ultimately seeking to return to the inorganic, he offers a resolutely materialist perspective on human origins and ends. Reproductive technologies like cloning confront us with a technologized version of this scene of origin, as when Iris visits the lab where she was made and sees embryos being created in 'a small cylindrical container, connected to a computer. Something was churning or growing in the vat; on the screen, amoeboid shapes split' (94).[14] In response, she is overcome by the feeling that within her there is 'another kind of Being, inorganic, non-biological, non-human entirely. The Weirdness. The Thing. The black matter lurking in the back of myself, into which I could vanish or metamorphose' (13). The experience compels an awareness of the materiality and contingency of origins and an understanding of the gap between the animate and the inanimate as being both immeasurably large and vanishingly small. In this respect, the novel deconstructs what Jackie Stacey calls the| 'bio-aura' which surrounds traditional forms of reproduction. Drawing on Walter Benjamin's account of the loss of the aura of the work of art in an age of mechanical reproduction, Stacey argues that genetic manipulation can be thought of as the biological equivalent to the shock of modernity discussed by Benjamin, as it threatens the 'sense of humanness 'associated with 'a particular generational capacity'.[15] It is notoriously difficult to determine whether Benjamin's aura refers to an illusion or reality, and this leads Stacey to suggest that the sense of aura is defined precisely by its loss, which produces 'an illusory or mythical past' (187). Iris's trajectory in the

---

[14] Hoffman makes this point in an interview when she comments that 'we always walk a delicate line between feeling ourselves to be mysterious and in some way absolute, and at the same time knowing we are material, we came from these particular people. The idea of cloning pushes this further'. See Brenda Webster, 'Conversation with Eva Hoffman', *Women's Studies: An Interdisciplinary Journal* 32 (2003), 761–9 (764), published online 29 Oct. 2010, doi: 10.1080/00497870390221927.

[15] Jackie Stacey, *The Cinematic Life of the Gene* (Durham and London: Duke University Press, 2010), p. 184. Further references will be given within the main text.

novel can be understood in terms of a growing recognition that the idea of a 'natural' genealogy may be just such an illusion, not least because of the long history of medical intervention in pregnancy and childbirth. *The Secret* also explores the idea that cloning, as image and practice, will dislodge sexual difference as the cornerstone of kinship and the social order. In this respect, it develops perspectives which intersect with Catherine Malabou's compelling reading of cloning in her essay 'Following Generations', which takes Claude Lévi Strauss's reading of Guillaume Apollinaire's poem 'Autumn Crocuses' as its starting point. As Malabou points out, Lévi Strauss's reading of the poem focuses on Apollinaire's enigmatic description of the autumn crocuses as 'like their mothers / Daughters of their daughters' and traces its significance in three distinct but overlapping registers.[16] From a botanical point of view, the autumn crocus represents 'a kind of vegetable time bomb': the flowers appear before its leaves and the leaves appear before the seeds, so that the order of secession is inverted and 'the after comes before the before' (24). However, the seeds play only a 'circumstantial' role in generation, as the crocus is a hermaphrodite which reproduces itself underground, so that the phrase 'mothers / Daughters of their daughters' also refers to the combining of two modes of reproduction, one vertical and one horizontal. From a mystical point of view, the autumn crocus is aligned with the Virgin who is both mother and daughter of God, and with ancient divinities which regenerate themselves without ageing: in this sense, it represents a twinning of the principles of generation and regeneration (28). Finally, in the symbolic register, the reproductive dualism of the crocus can be read as a figure for the reversibility of signifier and signified. For the structural anthropologist Lévi Strauss, therefore, the crocus corresponds to no existing linguistic or kinship structure but is an augury of 'a structure to come' (21).

In the novel, cloning is transposed from the natural to the techno-logical realm and associated with privileged figures such as Iris's mother Elizabeth, who has had an extremely successful career as an investment consultant. It is also linked with female agency, as Hoffman depicts a twenty-first-century society in which men have 'lost their adventure' while women have gained access to education and careers

---

[16] Catherine Malabou, 'Following Generation', trans. Simon Porzak, *Qui Parle* 20: 2 (Spring/Summer 2012), 19–33 (24). Further references will be given within the main text.

that were unavailable to them in the past. As women have amassed educational and financial capital, traditional family bonds have loosened, single parents have become the norm, and the nuclear family has faded into the past. It survives only in figures like Iris's grandparents, who are pillars of their 'elder community' and embody a social order grounded in the binary logic of sex and the paternal law (Iris's grandfather is 'a big man in every way') (132). By placing cloning in the context of these transitions, Hoffman emphasizes the fact that biology is never prior to culture but is produced and mediated by specific social conditions. Through carefully chosen references to anthropology, linguistics, and psychoanalysis she also draws attention to the way in which our beliefs about biology, kinship, and society reinforce each other. For much of the twentieth century, ideas about reproduction and kinship were shaped by the structuralist view of language as a system of binary oppositions. As meaning-making activities, reproduction and kinship were considered analogous to linguistic structures and were parsed in terms of binary oppositions (male/female; patrilineal/matrilineal descent). Even the unconscious was thought to be structured according to this binary logic, as Jacques Lacan claimed that the 'reality of the unconscious is sexual'.[17] The Secret maps what happens when this logic is undone, tracking the implications through Iris's development.[18] One way of interpreting her trajectory would be in terms of a protracted Oedipus complex. In her childhood she lives in a sealed bubble of closeness with her mother: they exist as if in 'a semi-liquid surround, an amniotic fluid that incorporated us both and within which there was a connecting passage or cord, along which silent sounds and messages and electrical pulses travelled back and forth' (16).[19] However, when Iris is

---

[17] Jacques Lacan, Four Fundamental Concepts in Psychoanalysis (1973), trans. Alan Sheridan (London: Routledge, 2018), p. 150.

[18] In this respect, as Hoffman points out in interview, she is staging a debate between science and psychoanalysis, a debate that she feels is overdue as 'psychoanalysis hasn't absorbed the new possibilities inherent in the genetic revolution' (Webster, 'Conversation', 766).

[19] The novel is informed by the extensive body of work in second-wave feminism which explores the early relationship between mothers and daughters, often focusing on the tensions arising from the mirroring relationship between them. See Nancy Chodorow, The Reproduction of Mothering: Psychoanalysis and the Sociology of Gender, second revised edition (Berkeley: University of California Press, 1999); Luce Irigaray, Speculum of the Other Woman, trans. Gillian C. Gill (Ithaca: Cornell University Press, 1985; Julia

twelve her mother has an affair with Steven Lontano, an archaeologist who studies, among other things, the incest taboo. We might expect this to precipitate an Oedipal crisis in which Iris would perceive her mother as lacking/castrated and would engage in a rivalrous struggle for the father-figure. However, for Iris the situation is complicated, for as she explains 'I didn't want to take Steven away from my mother; I wanted to share him with her. Or her with him. Or rather, I did share her—and him' (32). What is at stake is the oppositional logic of the Oedipal imaginary, which requires the child to line up on one side or other of a binary divide. This Oedipal logic both reflects and is legitimated by the two-into-one logic of sexual reproduction, whereby in meiosis the sex cells lose half of their chromosomes prior to forming a new organism, just as in the passage through the Oedipus complex, the child must relinquish and/or repress their desire for one of their parents. Iris refuses this choice, not from a tendency to infantile regression but in order to forge forms of filiation which reflect her non-normative origins. She rejects the primacy of the law of the father articulated by Steven ('Step-father's law'), wanting him to play a more peripheral part in family life, in line with the 'circumstantial' nature of the part played by sexual repro-duction in her origin (34). Her primary affiliation remains with her mother, who after the cloning revelation comes to signify not Freudian lack but its converse, plenitude. Like the autumn crocus in Malabou's reading, Elizabeth signifies 'a reproductive dualism that seems, through its excess, to bring reproduction to a halt'. By combining two types of reproduction in a single body, she dislodges sexual reproduction from its place as the foundation of the biosocial order and in this sense represents 'the very deconstruction of structures of kinship' which Malabou reads between the lines of Lévi Strauss's essay. In this context, Iris's continued attachment to her can be interpreted as post- rather than pre-Oedipal, as she affiliates with her mother/twin as much as she does with her lover

Kristeva, *Black Sun: Depression and Melancholia* (New York: Columbia University Press, 1992). More recently, the artist, psychologist, and psychoanalyst Bracha Ettinger has developed the influential concept of the matrixial, which offers a more positive interpret-ation of the trans-subjectivity of early life and of the relationships between mothers and daughters. See Bracha Ettinger, *The Matrixial Gaze* (Minneapolis: University of Minnesota Press, 2006).

Robert, simultaneously disturbing the binary logic of kinship and the linear logic of generation.

In addition to examining the implications of cloning for kinship, the novel registers the ways in which cloning effects a transformation in our understanding of living systems. Malabou's work is again pertinent in this respect, as she points out in 'Following Generation' that far from being the pure and simple replica predicted by neo-Darwinian theory, the clone is constituted by processes of de- and re-differentiation which transform our understanding of difference itself. As she puts it:

> the fact that it's possible to become daughter of your daughter, to be simultaneously older and younger than yourself [...] produces difference not in the sense to which we've become accustomed by good old DNA and other such kinds of code—a difference between individuals—but a difference *between code and message*. What allows us to think about the transfer of the nucleus in cloning is not so much the repetitive and stereotyped character of the results of this transfer, as scientific doxa repeats over and over again, but the fact that the nucleus is, precisely, transferable, displaceable, that birth can follow from a demobilization, from *a non-disseminatory difference between the nucleus and itself.* (32–3)

For Malabou we must expand our understanding of difference, taking it beyond the neo-Darwinian logic according to which the genetic programme specifies individual identities. We should see the structure of the living in terms of an intersection between the given and the constructed which means that it is impossible to establish a strict border 'between natural necessity and self-invention'. Hoffman's novel follows this line of thought, as Iris moves away from an understanding of herself as 'a replica, an artificial mechanism' (61). Rather than being a copy of her mother she discovers that her identity is articulated through recursive interactions which create a 'beginningless, end-less, self-perpetuating loop. A loop along which signals travelled, back and forth' (240). In addition, as she studies Bio-Theory and the forms of sub-cellular life, she comes to realize that every time she thinks she understands their patterns, this only prompts 'new, mind dizzying questions. For it was all too perfect, too arbitrary, too coordinated and too adventitious for comparison' (261). This cluster of oppositions suggests a perspective on difference which is close to

Malabou's parsing of difference in terms of plasticity, that is, an interplay between form and formlessness, essence and accident, in an economy of ceaseless change. The novel also explores the implications of cloning for our understanding of biological time. As Marilyn Strathern and Sarah Franklin have pointed out, Darwinian theory brought about a transformation whereby 'natural time' was reconceptualized in terms of biological development.[20] As linear time was biologized it was linked both with adaptive variation and with the irreversible sequence of generations, a perspective which, as Franklin suggests, maps on to a broader modern conception of time as 'one-way and finite in relation to the past, but also capable of signifying open-ended and multiple possibilities in the future'. However, developments in reproductive technology, especially cloning, have changed our understanding of biological temporality. The discovery that cellular differentiation can be reversed opens up the prospect of rewinding biological time to create 'more' time, either creating life through reproductive cloning or extending it through therapeutic cloning. The novel focuses on the way in which reproductive cloning might scramble the generations, creating a kind of warp in time. On the one hand, it can short circuit the temporality of generations, as we see in Iris's relationship with her grandmother, whom she sees as in some sense her mother, while Edith confuses her with her daughter, Elizabeth. Edith's confusion is related not only to the physical similarity between Iris and Elizabeth but to age-related memory loss, so that here the rewinding of biological time is linked to degeneration. On the other hand, through the relationship between Iris and her aunt Janey, Hoffman suggests the regenerative potential of such a warp in time. For Janey, Iris is not only her niece but is also a 'reincarnated sister' whose appearance gives her a chance to 'redress the past' (199). Iris's similarity to Elizabeth triggers visceral memories which allow Janey to 'slip back' in time and recognize the complexity and ambivalence of her feelings about her sister. In this case, going backwards in time restores the multiplicity and open-endedness of the past, enabling

---

See Marilyn Strathern, *After Nature: English Kinship in the Late Twentieth Century* (Cambridge: Cambridge University Press, 1992), chapter 1; Sarah Franklin, 'Rethinking Reproductive Politics in Time, and Time in UK Reproductive Politics: 1978–2008', *Journal of the Royal Anthropological Institute (NS)* (2014), 109–25 (109–10).

a recalibration of the relationship between the sisters. The novel thus demonstrates the plasticity of biological time in a dual sense. As the relations between the generations are reconfigured, biological time can appear to be compressed (as it is for Edith) or in some sense expanded (as it is for Janey). In addition, as a discursive category, biological time is revealed to be not natural and self-evident but a mutable socio-political formation.

For Hoffman the figure of the clone is a prism through which to refract the biopolitics of the past and the future. On the one hand, the novel looks back to the Holocaust and the eugenetic thinking which was imbricated with it, a point emphasized by Giorgio Agamben in *Homo Sacer* when he argues that genetic concepts did not have to be distorted by the Nazis but were ready-formed for their purposes: 'Nazism, contrary to a common prejudice, did not limit itself to using and twisting biological concepts for its own ends. The relationship between National Socialist ideology and the social and biological sciences of the time—in particular genetics—is more intimate and complex, and, at the same time, more disturbing.'[21] While these are bold claims, they are supported by the historians of science Staffan Muller-Wille and Hans-Jorg Rheinberger when they argue that classical genetics 'centered around questions of eugenics, racial identity and sexuality—in short a biopolitics of what came to be called the "racial body". This biopolitics culminated in World War 2 and the Holocaust.'[22] Hoffman's novel reflects on the long aftermath of this biopolitics while drawing out the transformative potential of the 'new biology' which was emerging at the turn of the millennium, and maps alternative modes of filiation with the potential to re-form hierarchical kinship structures. Ishiguro's *Never Let Me Go* is also informed by twentieth-century biopolitics but mobilizes cloning as a metaphor for more contemporary forms of biopolitical oppression. The novel is premised on the success of a line of research which had been more or less abandoned by the time when it was written, so that as Ishiguro notes, it posits 'an "alternative history" situation. The Illmensee controversies, etc, in the 70s, went the other

---

[21] Giorgio Agamben, *Homo Sacer: Sovereign Power and Bare Life*, trans. Daniel Heller-Roazen (1995) (Stanford: Stanford University Press, 1998), pp. 145–6.

[22] Stafan Müller-Wille and Hans-Jörg Rheinberger, *A Cultural History of Heredity* (Chicago and London: Chicago University Press, 2012), p. 185.

way: cloning, instead of getting set back at that period, went full steam ahead.'[23] As Gina Kolata explains in her book *Clone* (which Ishiguro read when he was drafting the novel), the biologist Karl Illmensee claimed in 1979 that he had cloned three mice from mouse embryos, in an intricate and difficult experiment. However, in 1983 he was accused of falsifying his results and the scandal tainted the field of cloning research: as Kolata writes, scientists not only turned their backs on Illmensee but 'began to disdain the very pursuit, and funding, of cloning research' (133). Moreover, by the 1990s research into tissue engineering was taking off, with successes in creating replacement skin, bladder, and other tissue, rendering whole body cloning redundant as a therapeutic technique. Nonetheless, as a metaphor, the full body clone operates as a powerful vector for the novel's biopolitical concerns.

The effects of the wars and genocides of the twentieth century reverberate across Ishiguro's fiction: in his first novel, *A Pale View of Hills*, a middle-aged Japanese woman reflects on her traumatic experience of motherhood in the wake of the destruction of Nagasaki, while his second, *An Artist of the Floating World*, assesses the culpability—or otherwise—of a Japanese painter who produces right-wing propaganda immediately prior to the Second World War. In similar vein, *When We Were Orphans* examines the entanglement of Japanese imperialism and British colonial power in Shanghai during the Sino-Japanese war, while *The Remains of the Day* explores the eugenic logic that found its most extreme expression in the Holocaust by tracing the Fascist sympathies of the British upper-classes in the run-up to World War Two.[24] Ishiguro's fictional example of this tendency is Lord Darlington, a character loosely modelled on Lord Halifax, one of the architects of appeasement whose memoir is an attempt to manage public memory of British pro-Nazism in the 1930s.[25] In his notes for the novel, Ishiguro emphasizes how mainstream such sympathies were among the upper classes,

---

[23] Box 5.4 Ishiguro Never Let Me Go: 'Ideas as they Come', Notebook 1, Jan '01–Sept '02, 22/2/01, Kazuo Ishiguro Archive, Harry Ransom Center, University of Texas at Austin.

[24] Ishiguro's other two novels, *The Unconsoled* and *The Buried Giant*, refer more obliquely to the aftermath of political and cultural trauma.

[25] The title of the novel echoes that of Lord Halifax's memoir, *Fullness of Days* (1957). and many of the protagonists in the memoir also appear in the novel, for example Neville Chamberlain, Hitler, and Joachim von Ribbentrop.

noting that approval of Nazism went 'deep into the upper classes […] with at one pt the king of England an enthusiast'.[26] Lord Darlington turns out to be Hitler's single most useful pawn, yet the narrator, his butler Stevens, never questions his employer's behaviour. This is the result of his over-identification with his job and repression of any moral intuitions which might conflict with the demands of his employer, as evidenced when Lord Darlington asks him to dismiss two Jewish maids simply on the grounds that they are Jewish. When Stevens is asked if it has occurred to him that this would be wrong, he insists that 'our professional duty is not to our own foibles and sentiments, but to the wishes of our employer'. If Lord Darlington's country house is a microcosm of the pre-war British state, then Stevens' handling of this incident reveals the ease with which, in the context of a liberal democracy, an identification with one's role and social position can translate into an abnegation of moral responsibility.

The Holocaust and twentieth-century genocide also inform *Never Let Me Go*, surfacing in understated allusions, but the novel's central focus is on the global biopolitics which were emerging in the 1990s and which have continued to shape the twenty-first century. To illuminate the biopolitical concerns of the novel this chapter draws on the work of Giorgio Agamben, particularly *Homo Sacer* and *The Open*. Agamben's approach to biopolitics differs significantly from that of Foucault, as while Foucault sidesteps the question of sovereignty and focuses on the specific mechanisms and techniques of biopower, Agamben locates biopower at the heart of sovereignty. In so doing, he develops an interpretive model which has been (rightly) criticized as top-down and universalizing, failing to capture historically specific power relations and eliding differences of race, gender, and sexuality.[27]

---

[26] Box 17.3 Ishiguro The Remains of the Day—'Butler notes and "ideas as they come"' 1986–1987, Kazuo Ishiguro Archive, Harry Ransom Center, University of Texas at Austin.

[27] Catherine Mills, for example, notes the general neglect of gender and exclusion of women from Agamben's philosophical lexicon and comments with dismay on his account of pornography in *Profanations*. See Catherine Mills, *The Philosophy of Agamben* (London: Routledge, 2008), pp. 115, 136. In relation to issues of race and colonization, Stephen Morton argues that Agamben's account of the state of exception overlooks 'the historical experience of violent forms of colonial sovereignty and emergency from the standpoint of the colonised'. See Stephen Morton, *States of Emergency: Colonialism, Literature and Law* (Liverpool: Liverpool University Press, 2013), pp. 5–6.

Nonetheless, his work has captured the reach and pervasiveness of biopolitics in an era of globalization and the associated collapse of the nation-state and its juridical structures. It has been seen as prescient in relation to twenty-first-century abuses of state power, and Agamben's concept of the 'state of exception' has been widely used to critique such tactics as the emergency measures introduced by George W Bush after the 9/11 terrorist attacks. In developing this concept, Agamben looks back to the origins of Western political thought, focusing on the image of *homo sacer*, an obscure figure of archaic Roman law, and drawing on Aristotle's distinction between *zoe*, the 'fact of living common to all living being' and *bios*, 'the way of living proper to an individual or group'. *Homo sacer* is a figure who has committed a crime for which he is banished from society, with all his rights as a citizen being revoked: he can be killed with impunity but cannot be sacrificed in ritual form. He is thus included in the juridical order solely in the form of an exclusion, and it is this state of exception, this inclusion/exclusion, which for Agamben is the founding act of the Western political tradition. It allows sovereignty to distinguish between those lives worth living and those which are consigned to the category of what he calls 'bare life', which is neither *zoe* nor *bios* but is produced through an indistinction between the two. For Agamben, the most extreme instance of a state of exception is that which facilitated the Nazi concentration camps, which he calls 'the most absolute *condititio inhumana* that has ever existed on earth' (166). He traces the emergence of the juridico-political structure which made this possible through the states of exception declared by the Weimar governments from 1919–24 and the Nazi 'decree for the protection of the people' which suspended the rights of personal liberty and freedom of expression in 1933. Crucially, in this last piece of legislation, the prelude to the incarceration and murder of the Jews, no mention was made of the state of exception as being due to an external or factual danger. In consequence, Agamben argues, the state of exception became 'confused with juridical rule itself', making the concentration camp 'the most absolute biopolitical space ever to have been realized, in which power confronts nothing but pure life, without any mediation' (170–1). Yet the concentration camp is not an anomaly but the limit case of the law's capacity to act with reference only to itself. According to Agamben, we are all at increasing risk of being placed in the category of bare life, hence his controversial claim that the camp is the 'hidden matrix' of modernity (166). In support of this argument, Agamben draws attention to the millions of refugees who are detained in camps around the world without rights or representation, as are those held at camps such as Guantanamo Bay as terrorism suspects. He points out that such camps are not the only sites where bodies are managed as bare life: in the US in the wake of 9/11, the protection of the civilian population has entailed ever more intrusive procedures of surveillance and monitoring. In this respect, there is a dialectical relationship between the bare life which is excluded from the polity and the polity construed as bare life.

Never Let Me Go takes up the question of bare life in the figure of clones who are created expressly for the purpose of providing body parts for others. Although Ishiguro has insisted that the novel is 'essentially not about genetics and ethics, it's about the human condition', the metaphor of the clone inevitably broaches questions about the management of life in what Ian Wilmut calls 'the age of biological control' (24).[28] The novel posits a cloning programme which has become well established by the 1990s (when the novel is set) but which takes place in the shadows because it entails whole body cloning for organ donation and 'people preferred to believe these organs appeared from nowhere, or at most that they grew in a kind of vacuum'.[29] When the public is forced to confront the existence of the clones, they persuade themselves that the clones are not 'properly human' because this makes it easier to think that their deaths do not signify; moreover, if the clones are thought of as not really human, the frightening idea that cloning might be mobilized to create 'superior' children seems less plausible. Accordingly, the clones are segregated from wider society and educated in private boarding schools. Some, like Hailsham, where the narrator Kathy H grows up, are humane but most resemble the prison camps of the Second World War, being surrounded by electric fences (77). Subsequently, the clones move to college life in a remote setting where they begin to consider their

---

[28] See Box 5.6 Ishiguro Never Let me Go; 'First Rough Draft' Notebook—Clones 1' 2000–2001, 6/3/00, p. 5, Kazuo Ishiguro Archives, Harry Ransom Center, University of Texas at Austin.

[29] Kazuo Ishiguro, *Never Let Me Go* (2005) (London: Faber and Faber Limited, 2006), p. 257. Further references will be given within the main text.

future as carers and donors, while their last months are spent in euphemistically named 'recovery centres' where they die after their fourth donation.[30] In their marginalized and peripheral existence they resemble the refugees and asylum seekers who came to Britain in the 1990s from the former Yugoslavia, especially Bosnia and Kosovo, fleeing ethnic persecution. In this respect, the novel can be viewed as a reflection on human rights in a post-nation-state world, one which concentrates on the interrelation between refugees and the 'host' society.[31]

For Agamben, the refugee, 'the figure that should have incarnated the rights of man *par excellence*' now reveals 'the radical crisis of this concept'.[32] Drawing on the work of Hannah Arendt, he argues that the belief that non-citizens nonetheless have human rights, implicit in human rights declarations from the French Revolution to the twentieth century, no longer holds.[33] According to Arendt, the Holocaust had shown that 'the mere fact of being human' had become detached from the concept of rights and for Agamben, this process has only been amplified as the West has responded to mass migration by calling in question the rights of refugees. It has become clear that human rights are not portable, as asylum seekers are held in detention centres for

---

[30] The recovery centre where Kathy cares for Tommy is particularly resonant in this respect, being a holiday camp 'for ordinary families' which has been half-converted, so that the clones are living, or rather dying, amidst an architecture which is sinister in its aspect of the management of the masses, while it is also poignant in its evocation of the pleasures of family life ('happy people [...] having a great time') which are denied to the clones (p. 214).

[31] The terms refugee, asylum seeker, and migrant are often confused. A refugee is a someone who 'owing to a well-founded fear of being persecuted for reasons of race, religion, nationality, membership of a particular social group or political opinion, is outside the country of his nationality' and is unable to return (UN Convention on Refugees). An asylum seeker is someone who has applied for refugee status but whose application has not yet been processed, while a migrant is someone who has moved for economic reasons. The terms asylum seeker and migrant have, quite unwarrantably, taken on negative connotations in the UK in recent years. For further details see https://www.amnesty.org.au/refugee-and-an-asylum-seeker-difference/ [accessed 14 April 2019].

[32] Giorgio Agamben, 'We Refugees', translated by Michael Rocke, *Symposium* 49, Summer 1995, Periodicals Archive Online 116, rhttps://thehubedu-production.s3.amazonaws.com/uploads/1836/1e78843In%20tho-c11e-4036-8251-5406847cd504/AgambenWeRefugees.pdf [accessed 14 April 2019].

[33] The source text for Agamben's essay is Hannah Arendt's chapter 'The Decline of the Nation-State and the End of the Rights of Man' in *The Origins of Totalitarianism* (New York: Meridian Books, 1958). Arendt was herself stateless and as she put it 'rightless' for seventeen years after leaving Nazi Germany in 1933.

indefinite periods and, if granted refugee status, are expected to take up the work that no one else wants. Yet refugees and asylum seekers are not outside society but live in a state of inclusion-exclusion, as do the clones in *Never Let Me Go*. They too inhabit a state of exception, included in the law by virtue of being excluded from it. They lack the rights associated with citizenship because they are not born 'naturally' to a nation, and their reproductive future has been removed because they are engineered to be sterile. As Anne Whitehead argues, their confrontation with their guardians at the end of the novel, when they attempt to defer their donations, is reminiscent in its staging of a legal trial or appeal but as such it merely underscores the fact that they 'have no access to the anonymous "they" who determine the narrow confines within which their lives can be lived'.[34] The parallel between the clones and refugees is also suggested by their employment as carers. In the West, refugees are disproportionately represented in the 'caring professions', as they often have no choice but to tolerate the associated low pay, lack of benefits, and antisocial hours. At the same time, they provide care which goes beyond the physical to include affective labour, as is the case for clones like Kathy, who provides both emotional and physical support for her donors: 'I have a kind of instinct around [them]. I know when to hang around and comfort them, when to leave them to themselves.' However, there is an asymmetry between the care the clones give and the care they receive from society which takes us to the heart of the novel's biopolitical concerns.[35]

To elucidate the nature of bare life, both Ishiguro and Agamben focus on the coma patient and their uneasy proximity to what Agamben calls the 'neomort', that is, the brain-dead patient waiting to donate their organs. As he argues in *Homo Sacer*, this proximity is the result of the convergence of the biomedical technologies of life support and of organ transplantation, which together create 'a space of exception in which a purely bare life, entirely controlled by man and his technology, has appeared for the first time' (164). In this zone of indistinction, the

---

[34] Anne Whitehead, 'Writing with Care: Kazuo Ishiguro's *Never Let Me Go*', *Contemporary Literature* 52 (Spring 2011), 54–83 (67).

[35] For a discussion of this asymmetry, and of the rights of refugees, asylum seekers, and economic migrants to care, see Eva Feder Kittay, 'The Moral Harm of Migrant Carework: Realizing a Global Right to Care', *Philosophical Topics* 371 (Spring 2009), 53–73.

border between life and death is indeterminate, hence the development of the concept of brain death, which Agamben sees as a biopolitical move in that it can allow for the transplantation of organs from a coma patient. The risks of relying on such criteria are illustrated by the case of Karen Quinlan, an American woman who was kept alive in a deep coma before her life support was switched off, at which point she started to breathe again and survived in a state of artificial nutrition for several years. Her ability to breathe naturally could not have been predicted, raising questions about what other capacities may be latent in coma patients.[36] Although her living body had become 'entirely separated from the form of life that bore [her] name', Quinlan's life is preserved, not as individual life but as an instance of life as a substance to be managed. In this respect, she can be aligned with the population which politicians seek to protect when they impose states of emergency.

However, a point Agamben does not make is that Quinlan is a white, middle-class American, a fact which may have played a part in the sacralization of her life. In contrast, the assignment of Ishiguro's clones to a state of living death is facilitated by the fact that, unlike Quinlan, they are understood to have come from 'trash'. As Ruth puts it in a discussion about the clones' parents, they have been cloned from 'junkies, prostitutes, winos, tramps. Convicts, maybe, just so long as they aren't psychos' (164). In addition to being linked with the addicted and dispossessed, they are derogated on the grounds of their supposed genetic difference and in this respect, the novel speaks to its 'post-racial' moment, the paradoxical logic of which has been analysed by Alys Weinbaum. As she suggests, the current view is that race has no genetic basis: 'ever since the announcement of the completion of the map of the human genome in June 2000, the case against race more often than not

---

[36] These issues are explored by the neuroscientist Adrian Owen in *Into the Grey Zone: A Neuroscientist Explains the Mysteries of the Brain and the Borders between Life and Death* (London: Guardian Faber Publishing, 2017). As he explains, the borders of consciousness are constantly shifting as some patients categorized as being in a persistent vegetative state (PVS) or a minimally conscious state (MCS) are found to have some form of awareness, as do patients with locked-in syndrome. As the neurosurgeon Henry Marsh notes in his review of Owen's book, the difficulty is knowing how the clinician should respond to this knowledge. See Henry Marsh, 'Into the Grey Zone: can we really be conscious while in a coma?', *New Statesman* 27 August 2017, https://www.newstatesman.com/culture/books/2017/08/grey-zone-can-one-really-be-conscious-while-coma [accessed 14 April 2019].

is presented in genetic terms and as definitively closed' (208). Nevertheless, this 'official' consensus coexists with 'a culture that continues to renew its commitment to the idea of race through its practice of biotechnology', for example in race-based medicine.[37] In consequence, in a supposedly post-racial age 'even genomic art without overt racial content is paradoxically haunted by racial aura' (226). Ishiguro's novel has no overt racial content but simply because of their alleged genetic difference, the clones are imbued with a racial aura which encourages their identification as not 'properly human' (258).[38] Their liminal status leads to fears about what happens after their fourth donation, fears which are revealed by Tommy when he explains that donors cannot be certain that they will 'complete', that is, die, at that point. There are rumours that:

> after the fourth donation, even if you've technically completed, you're still conscious in some sort of way; how then you find there are more donations, plenty of them, on the other side of that line; how there are no more recovery centres, no carers, no friends; how there's nothing to do except watch your remaining donations until they switch you off. It's horror movie stuff, and most of the time people don't want to think about it. (274)[39]

Through this extreme instance of biopolitical appropriation, in which the potential for a residue of consciousness is brutally discounted, Ishiguro underscores the broader disregard of the clones' subjectivity, a point which is also made by Nancy Armstrong when she points out

---

[37] Alys Eve Weinbaum, 'Racial Aura: Walter Benjamin and the Work of Art in a Biotechnological Age', *Literature and Medicine* 26, Spring 2007, 207–39 (208, 226). For an excellent analysis of the figuration of race in *Never Let Me Go*, see Josie Gill, 'Written on the Face: Race and Expression in Kazuo Ishiguro's *Never Let Me Go*', *Modern Fiction Studies* 60, Winter 2014, 844–62.

[38] Ishiguro's notes for the novel are suggestive in this respect, as he writes in relation to the clones that the book should have a 'multi-ethnic dimension [. . .] though we don't make a big deal of it. A descriptive thing . . . but yes, a lot of them would be mixed genetically'. Box 5.4 Ishiguro Never Let Me Go: 'Ideas as they Come', Notebook 1, Jan '01–Sept '02, 14/1/02, Kazuo Ishiguro Archive, Harry Ransom Center, University of Texas at Austin.

[39] Ishiguro's use of the term 'complete' echoes the medical use of the term in relation to those w1ho commit suicide (suicide completers). Its use has been seen as unhelpful in the medical context in that it implies the successful carrying out of a project, but this is exactly the sense that Ishiguro intends.

that the relations between clones and non-clones are marked by a stark absence of intersubjectivity, the clones being relegated to the status of 'third persons who are by their very nature ineligible for personhood'.[40] The question of the clones' personhood is explored through the debate about their education, which echoes the debates about social engineering sparked by evolutionary psychology in the 1990s. On the one hand are those who believe that by virtue of their genetic difference, the clones are nonhumans who are not worth the effort of a civilized education. In general, society is happy to let them be reared in 'deplorable conditions' but this practice is challenged by a small group of progressives who dedicate themselves to showing that 'if students were reared in humane, cultivated environments, it was possible for them to grow up to be as sensitive and intelligent as any ordinary human being' (256). While Ishiguro does not overstress the point, such an experiment in 'humane' education can be aligned with the constructionist view, derided by evolutionary psychologists, that human behaviour is malleable and is primarily shaped by the environment rather than by genes.[41] The education the clones receive at Hailsham is humane and humanist, being premised on the Romantic view, derived from Immanuel Kant and Friedrich Schiller, that a sense of the beautiful is a defining characteristic of the human being. Accordingly, the clones are brought up in beautiful surroundings and are encouraged to create poems, paintings, and sculpture which are traded between them at monthly 'Exchanges', while the best items are removed and shown in public in order to prove that the clones have souls. In this view, creativity is definitive of the human being, who is separated from the animal by the possession of a soul. However, the clones' aesthetic education raises the fundamental question which Kathy puts to the 'Guardians' (i.e. the teachers) in the 'trial' scene: 'Why train us, encourage us, make us produce all of that? If we're just going to give donations anyway, then die, why all those lessons?' (254). An art based on human exceptionalism does not meet

[40] Nancy Armstrong, 'The Affective Turn in Contemporary Fiction', *Contemporary Literature* 55 (Fall 2014), 441–65 (448).
[41] Ishiguro makes this point in his notes, writing that 'Madame, amongst other beliefs (see previous) believed that you could improve human nature by changing environment—the old "left" dream'. Box 5.4 Ishiguro Never Let Me Go: 'Ideas as they Come', Notebook 1, Jan '01–Sept '02, 21/3/01, Kazuo Ishiguro Archive, Harry Ransom Center, University of Texas at Austin.

their case and cannot capture their situation, nor, the novel suggests, does it speak to the position of 'ordinary humans' like the Guardians. In this final scene, we see the impoverished Guardians selling off 'beautiful objects' from Hailsham with no regret: for them the value of such art has become purely economic.

The novel thus suggests that humanist art is inadequate to a contemporary moment in which the border between the human and the nonhuman has become problematic. However, it also points to limitations in the biological view of human existence offered by neo-Darwinism.[42] The neo-Darwinian perspective is represented in the novel at a point when Tommy, anxious to prove that he is creative and so worthy of a deferral, begins to sketch imaginary animals, taking up the role of evolutionary designer in this respect. When Kathy first sees his animals, they appear mechanical, robotic: 'the first impression was like one you'd get if you took the back off a radio set: tiny canals, weaving tendons, miniature screws and wheels were all drawn with obsessive precision' (185). However, when she looks more closely, she notices that there is 'something sweet and vulnerable' about them and recalls Tommy saying that 'he worried, even as he created them, how they'd protect themselves and be able to reach and fetch things' (184). Through this 'just so' evolutionary narrative, in which Tommy appears as a concerned parent-creator, the novel points to the anthropocentrism which inflects evolutionary theory, together with a human exceptionalism which entails the belief that, as Daniel Dennett puts it, 'there is a huge difference between our minds and the minds of other

---

[42] The novel's engagement with neo-Darwinian theory is indicated by comments in a small notebook among Ishiguro's papers, in which notes on 'genetic inheritance' and free will are followed by a page on which he has written 'Darwin's Dangerous Ideas'. Alongside this reference to Daniel Dennett's *Darwin's Dangerous Idea*—one of the most forceful accounts of neo-Darwinian theory—is a visual diagram of the territory over which the Darwin wars were fought. This maps a continuum of positions from anti-Darwin Creationism to ultra-Darwinism and marks three key thresholds: between Creationists and those who believe in a soul (the 'Darwin threshold'); between believers in a soul and materialists (the 'materialism threshold'); between those who believe that 'the environment is all' and those who believe that 'to understand us you must understand our evolution' (the 'EP threshold'). EP refers to evolutionary psychology. Box 3.1 Ishiguro Never Let Me Go—'Rough Papers' (1 of 15) Feb. 2001–July 2003, Kazuo Ishiguro Archive, Harry Ransom Center, University of Texas at Austin.

species, a gulf wide enough even to make a moral difference'.[43] Neo-Darwinism, in short, is limited by its scientific or secular humanism, its own construction of a border between the human and the nonhuman.[44]

In exploring the human–nonhuman border, Ishiguro takes a similar approach to Agamben, focusing on the threshold between the human and the animal. For both, what is in question is not the animal as a being other than us, with whom we might have relations of companionship, hostility, or mutual dependency. Rather, their concern is with human being as such, and with the border between man and animal as a division within the human. As Agamben argues in *The Open*, the category of the human is produced and maintained against the category of the animal, or more precisely, the human recognizes itself by making a distinction between its animalistic and humanistic dimensions. His aim is not to resolve or deconstruct this opposition but to reflect on the space of exception in which the terms are articulated, which is an empty zone in which the 'truly human being who should occur there is only the place of a ceaselessly updated decision'.[45] For him, *Homo sapiens* is 'neither a clearly defined species nor a substance; it is, rather, a machine or device for producing the recognition of the human', a device which he dubs the 'anthropological machine'. We see this machine in operation in the novel as the clones are animalized so that 'ordinary' humans can define themselves in opposition to them: for example, one of their guardians always avoids touching them and is afraid of them 'in the same way someone might be afraid of spiders' (35).

---

[43] Daniel C. Dennett, *Darwin's Dangerous Idea: Evolution and the Meanings of Life* (1995) (London: Penguin, 1996), p. 371.

[44] Scientific humanism is a term which was espoused by E.O. Wilson and Richard Dawkins in the 1970s and 1980s but the term secular humanism is now more often used to denote a commitment to rational enquiry through science and philosophy. A strand of human exceptionalism, particularly in relation to cognition and morality, still runs through the arguments of many secular humanists. See Dennett pp. 477–81 for a discussion of this.

[45] Giorgio Agamben, *The Open: Man and Animal*, trans. Kevin Attell (2002) (Stanford: Stanford University Press, 2004), p. 88. Further references will be given within the main text. Agamben's position on human exceptionalism is ambivalent. Unlike posthumanist thinkers such as Rosi Braidotti, he does not assert a continuity or isomorphism between the animal and the human, nor on the other hand does he insist on the distinction between the two. Rather, he suggests in *The Open* that we should learn to think of the human 'as what results from the incongruity of these two elements' (p. 16).

Conversely, when the clones are receiving their humanist education, they construct themselves as human by making paintings and sculptures of animal others. Tommy's watercolour of an elephant in the grass is exemplary in this respect, except that it is 'the kind of picture a kid three years younger might have done'. He is relentlessly teased about his naïve painting, which, however, can be seen to reveal the kind of potential which Agamben locates at the point when the distinction between human and animal breaks down. On one level, the painting is clearly an allusion to the 'elephant in the room' of the clone condition, expressing Tommy's intuition of the reality of their situation. On another, it represents a challenge to stylized representations of animals, which represent an idea rather than a relation, and an openness to the human–animal threshold. This perspective fades as Tommy progresses through the successive stages of his education, to the point where he creates animals made of metal and rubber which seem lifeless to Kathy. By this stage he is already making his donations and has been assimilated to the logic of what Agamben calls in *The Open* the 'total management' of the animality of man (77).

Agamben wants us to reject such biopolitical strategies and to recognize the enigmatic relation between the human and the animal. Unlike Heidegger, he does not dismiss the animal as 'poor in world' but argues that the animal is 'captivated' by an external world which is nonetheless concealed from it. In its absorption in the world, Agamben sees 'a more spellbinding and intense openness than any kind of human knowledge' (59). For humans, in contrast, the world is constituted as a series of meaningful projects, and it is only when something interrupts their flow that we become aware of the world in what Heidegger would call its 'thereness'. In such disruptions, which are prompted by boredom or angst, we become aware of the existence of a world apart from 'our' world and are aware of our own captivation, or as Agamben puts it, we 'remember captivation an instant before the world disclosed itself' (70) This awakening to our own being-captivated does not open onto 'a further, wider and brighter space', rather, 'whoever looks in the open sees only a closing'. In such moments, the human being is in the closest proximity to the animal because 'both are, in their most proper gesture, *open to a closedness*; they are totally delivered over to something that obstinately refuses itself' (65). The final scene of the novel depicts such a moment, as Kathy drives across the featureless

Norfolk landscape, disturbing the flocks of birds which are the only instance of animal life in the novel. The birds resemble the 'strange rubbish' which is blown across the fields, which is in turn identified with everything that Kathy has ever lost so that an equivalence is established between the 'captivated' flight of the birds and Kathy's progress through life. As she imagines Tommy, who has died two weeks earlier, coming towards her across the expanse of earth and sky, she perceives her own finitude and registers the closedness of the external world, with its 'flat fields of nothing'. The novel thus points to the way in which for clones as for 'ordinary humans', the threshold between the human and the animal is our inmost truth.

By focusing on the human–animal threshold, the novel also broaches the theme of indifference which, for Claire Colebrook and Jason Maxwell, resonates across Agamben's work.[46] As they point out, the importance of difference has been taken for granted in recent decades. At the level of theory, the concept of difference as origin has been central to the work of Derrida and Deleuze, while in wider society the right to be different has been linked to a celebration of diversity in all its forms. Most crucially, at the juridical level, difference is associated with the sovereign decision which installs the distinction between what is inside and what is outside the law. Agamben takes a different approach, challenging the view that everything begins with the differences that produce relations and directing attention to the indifference, or pre-relational zone, from which differences emerge. He argues that as beings who are poised on the threshold between speech and silence, humans come into being by fulfilling a potential which is always coupled with impotential, that is, with the possibility that the potential might not be fulfilled. As a corollary, the space of politics and citizenship in which rights and laws come into being might also always *not* be constituted. There is nothing inevitable about the current formation of the human as a political subject, which means that it might be possible to rethink human community and forge a politics which is not premised on the recognition of already-established identities. In this

---

[46] Claire Colebrook and Jason Maxwell, *Agamben*, Key Contemporary Thinkers (Cambridge: Polity Press, 2016). For them, 'it is the concept of difference and indifference that is most useful in recognizing the distinction and resonance of Agamben's corpus' (p. 195).

respect, Agamben envisages a community which is based on 'whatever being', that is, a life which is not yet identified and actualized as the subject of human rights.[47] This is neither bare life, nor the life of the political subject, but life in its singularity, the singularity of being here. In this vision, community is forged through what Heidegger would term a 'being with' which is both before and outside the relations which bind the subject to the polity.

In their introduction to a collection of essays on Ishiguro's work, Sean Matthews and Sebastian Groes argue that *Never Let Me Go* is haunted by the idea that we are not proper, rounded subjects, and that this leads to a sense of loss that can only be recuperated 'through forging, nursing and celebrating brittle human relationships': in this view, humanist values are salvaged even as their frailty is acknowledged.[48] Equally, it could be argued that the novel is concerned to map an alternative way of being in the world, one which is at an angle to both liberal humanist subjectivity and the abjection of bare life. As Armstrong points out, at Hailsham the clones run around in packs that give them little privacy and 'enjoy a kind of intimacy that is also profoundly social'. They perform a limited number of games and sexual behaviours for each other, participating in a community which is more like a composite body, joined by new members as others leave to 'complete' (458). The bonds between them are pre-individual, attuned to the singularity of being in the present, and this is nowhere more apparent than when donors and carers navigate the donations process. At this point, the donor is very far from being an individual subject defined by meaningful projects, as their future is foreclosed and they are confronted by finitude, while the carer knows that they will soon suffer the same fate. Nonetheless, when Kathy and Ruth spend time together before Ruth's final donation, they experience a 'being with' which eludes or precedes relations of identity and difference. As Kathy recounts, 'we'd sit side by side at her window, watching the sun go down over the roofs, talking about Hailsham, the Cottages, anything that drifted into our minds' (231). The sense of drifting signals a

---

[47] See Giorgio Agamben, *The Coming Community*, trans. Michael Hardt (1990) (Minneapolis: University of Minnesota Press, 1993), p. 17.

[48] Sean Matthews and Sebastian Groes, *Kazuo Ishiguro: Contemporary Critical Perspectives* (London: Continuum, 2010), p. 9.

relation attuned to the singularity of being and invokes the kind of community Agamben imagines, while also pointing to an ethics of attention which is not grounded in humanist conceptions of identity. From this point of view, the clones' compliance with their fate, which has frustrated many readers, can be read as a refusal to engage with a juridico-political system which excludes their own bare life, and as demonstrating a commitment to the recognition of 'whatever being' in the sense in which Agamben defines in in *The Coming Community*, that is, an understanding of 'being such that it always matters'.

For both Ishiguro and Hoffman, the trope of the clone prompts a critical examination of difference as the organizing principle of twentieth-century thought. The privileging of difference can be traced back to Ferdinand de Saussure's view of language as a system of binary oppositions, which informed structuralist approaches in anthropology and psychoanalysis, so that ideas about reproduction, kinship, and language reinforced each other in a circular fashion. As Hoffman's novel shows, the figure of the clone makes a dent in this logic, dislodging sexual reproduction from its primary place in the biological and social order and denaturalizing the link between heterosexual reproduction and kinship. Cloning technology has also transformed our understanding of living systems, demonstrating that cellular differentiation can be reversed and that the genome can be reactivated. As biological differentiation is defined in terms of interactive transformations, the neo-Darwinian genetics of fixed codes and structures is displaced by a model of dynamic and responsive self-differentiation. It is for this reason that Malabou argues in 'One Life Only' that biology might itself be a resource for contesting the biopolitical 'control, regulation, exploitation and instrumentalization' of the living being. For her, the biological is 'a complex and contradictory authority, opposed to itself and referring both to the ideological vehicle of modern sovereignty and to that which holds it in check' (1). Hoffman's novel replicates this insight, mobilizing the new biology to challenge genetic determinism and the idea of cloning as replication, while *Never Let Me Go* deploys the subjectivity of clones to challenge the biopolitical exclusion of bare life and the conception of the human that enables it. Like Agamben, Ishiguro underscores the link between ideas of genetic differentiation and the construction of bare life, and like him, he suggests that we might find an alternative to global biopolitics by attending to the fragile opening of life before it is either assimilated to sovereignty or derogated as bare life. This perspective entails a focus on the zone of indifference that precedes the forming of differences, and it also leads to an invocation of the pre-personal and the pre-political as a locus of resistance to the biopolitical. This could be viewed as a form of political quietism, a point which is not unrelated to the supposed compliance and passivity of the clones.[49] This is certainly a charge which has been levelled at Agamben by his critics.[50] However, such a reading of Ishiguro's novel is hard to sustain in the light of the radical force of its indictment of sovereign power, biopolitics, and the oppositional categories that shape our thought, an indictment accompanied by a suggestive invocation of alternative possibilities.

[49] In the *Guardian*'s Book Club weblog, according to John Mullan, the issue of the clones' failure to rebel provoked the most animated disputes. See John Mullan, 'Positive feedback', *Guardian*, Saturday 1 April 2006, available at https://www.theguardian.com/books/2006/apr/01/kazuoishiguro [accessed 14 April 2019].

[50] Antonio Negri sees Agamben's work as divided between a Marxist materialism and a Heideggerian nihilism, and in his view, the latter tends to prevail. As he writes in a review of *Opus Dei*, Agamben 'moves against any humanism, against any possibility of action, against any hope for revolution'. See Antonio Negri, 'The Sacred Dilemma of Inoperosity: On Giorgio Agamben's *Opus Dei*, Unimode 2.0, 18/09/2012, archived at http://www.uninomade.org/negri-on-agamben-opus-dei/ [accessed 14 April 2019]. On a related note, Rosi Braidotti has expressed concern about the fatalism which she finds in Agamben's philosophy, linking it to a 'fixation on *Thanatos* that Nietzsche criticized over a century ago [...] it often produces a gloomy and pessimistic vision not only of power, but also of the technological developments that propel the regimes of biopower'. See Rosi Bradotti, *The Posthuman* (Cambridge: Polity Press, 2013), p. 121.

| 5 |

Postgenomic Histories

Margaret Drabble and Jackie Kay

When the Human Genome Project (HGP) was completed in 2003, its findings confounded many of the expectations that had informed the initial planning. The first surprise was that the human genome contained far fewer genes than had been anticipated (around 25,000 as opposed to 80,000), and the second was that the proportion of human DNA that codes for protein is less than 2 per cent, raising the question of what the remaining 98 per cent is for. In addition, in 2008 genome-wide association studies (GWAS) identified the 'missing heritability' problem, that is, the fact that even when the genomes of thousands of individuals were sampled, genetic variation was unable to account for much of the inheritance of traits. It was clear that the workings of the genome were far more complicated than had been anticipated, prompting a surge of research in epigenetics which has shown the extent to which gene expression is modified in response to environmental cues and which has also demonstrated that many of these changes are heritable.[1] This has brought a shift from the atomism

---

[1] The issue of transgenerational epigenetic effects has been contentious, not least because of debates about the technical difference between intergenerational effects (occurring between two generations) and transgenerational effects (occurring across multiple generations). Nonetheless, the evidence for transgenerational epigenetic effects

Genetics and the Literary Imagination. Clare Hanson, Oxford University Press (2020). © Clare Hanson.
DOI: 10.1093/oso/9780198813286.001.0001

and determinism of neo-Darwinism to the opportunities and uncertainties of the more fluid postgenomic era. The term postgenomic can be used in a temporal sense to refer to the period after the sequencing of the human genome and it can also be defined technically, with reference to the advent of whole-genome technologies such as databases, biobanks, and microarray chips for assessing the expression of thousands of genes. More importantly for those working in the social sciences and the humanities, the postgenomic era can be seen to inaugurate a new style of biological thought, with epigenetics at its centre.[2] In place of the neo-Darwinian view of the sealed genome, epigenetics tracks the responses of the genome to changing environments, especially during critical windows of plasticity. As the genome is reconfigured as a molecular archive of responses to experience, epigenetics points to the intersection of history and biology. As Jorg Niewöhner puts it, epigenetics produces an 'embedded body, a body which is imprinted by its own past and by its social and maternal environment, and by evolutionary and transgenerational time'.[3] This chapter discusses two texts which work across the transition from neo-Darwinism to postgenomics, and which explore the ways in which experience, including the experience of previous generations, is 'written on the body', to use Jeanette Winterson's evocative phrase.[4] Margaret Drabble's *The Peppered Moth* was published in 2000, around the time of the completion of the first draft of the Human Genome Project, while

is compelling. A striking example comes from the UK Avon Longitudinal Study of Parents and Children (ALSPACH), which found an inverse correlation between the availability of food for men in childhood, and the longevity of their grandsons. See Pembrey, M.E., Bygren, L.O., Kaati, G., Edvinsson, S., Northstone, K., Sjöström, M., Golding. J.; ALSPAC Study Team, 'Sex-specific, male-line transgenerational responses in humans, *European Journal of Human Genetics* 14 (2006), 159–66.

[2] For a discussion of epigenetics as 'a different style of reasoning' see Maurizio Meloni, 'Race in an Epigenetic Time: Thinking Biology in the Plural', *British Journal of Sociology* 68 (September 2017), 389–409 (392), doi:10.1111/1468–4446.12248.

[3] Jorg Niewöhner, 'Epigenetics: Embedded Bodies and the Molecularisation of Biography and Milieu', *BioSocieties* 6, 2001, 279–98 (290), quoted in Sarah Richardson and Hallam Stevens, *Postgenomics* (Durham and London: Duke University Press, 2015), p. 226.

[4] See Jeanette Winterson, *Written on the Body* (London: Jonathan Cape, 1992). Foreshadowing epigenetics, she writes that 'written on the body is a secret code only visible in certain lights; the accumulations of a lifetime gather there' (p. 89).

Jackie Kay's *Red Dust Road* was published ten years later, just as epigenetics and other postgenomic perspectives were beginning to filter through to a non-specialist audience.[5] Both texts express unease with genetic determinism and probe the porous boundary between the biological and the social, intersecting strikingly with contemporary developments in the life sciences.

Drabble's novel is framed by the titular metaphor, which is usually seen as a textbook example of neo-Darwinian evolutionary theory. Peppered moths are normally white with black speckles across their wings, a pattern which provides camouflage against the bark of the birch trees that form their usual habitat. Following the Industrial Revolution, the air became blackened with soot in the north of England, darkening the tree trunks so that the light moths were picked off by predators. At this point, the dark form of the moth, which was better camouflaged against the blackened trees, began to outnumber the pale form to the point that the latter almost became extinct, until de-industrialization reversed the process. The cause of the change that gave the dark moth its adaptive advantage has long been debated, with the most frequent explanation being a random genetic mutation, in line with the neo-Darwinian view that evolution is driven solely by such mutations. However, the most recent research challenges this perspective, as in a 2016 paper Ilik J. Saccheri and colleagues showed that the mutation was caused by a so-called 'jumping gene', a transposon which had inserted itself into the *cortex* gene.[6] Transposons, which were first discovered by Barbara McLintock in 1948, are DNA sequences which can change position within the genome, creating or reversing mutations.[7] They have an important regulatory function as they can deactivate or otherwise alter gene expression and they intersect closely with epigenetic

[5] The best-known popular accounts of epigenetics are Nessa Carey, *The Epigenetics Revolution: How Modern Biology is Rewriting Our Understanding of Genetics, Disease and Inheritance* (London: Icon Books, 2012) and Tim Spector, *Identically Different: How You Can Change Your Genes* (London: Weidenfeld and Nicholson, 2012).

[6] See letter to *Nature*, 'The industrial melanism mutation in British peppered moths is a transposable element', Arjen E. van't Hof, Pascal Campagne, Daniel J. Rigden, Carl J. Yung, Jessica Lingley, Michael A. Quail, Neil Hall, Alistair C. Darby and Ilik J. Saccheri, *Nature* 534, 102–5 (02 June 2016), doi: 10.1038/nature17951.

[7] For an illuminating account of the neglect of Barbara McLintock's work see Evelyn Fox Keller, *A Feeling for the Organism: The Life and Work of Barbara McLintock* (1983) (New York: Holt Paperbacks, 2003).

processes. Research on transposons has contributed to the reconfiguration of the genome described above, whereby it is seen as a response mechanism which reprogrammes itself in response to environmental cues. However, neither the recent work on the peppered moth nor the wider postgenomic perspective was available to Drabble at the time of writing, so that the novel's probing of inheritance is framed by competing neo-Darwinian and neo-Lamarckian interpretive frameworks.

The Peppered Moth is set in the industrial north of England, but its focus is not on moths but on human adaptation to change. Specifically, the novel is concerned with the educational opportunities offered to the working classes in early twentieth-century Britain, bringing the promise of social mobility. The novel resembles D.H. Lawrence's *The Rainbow* (1915) in tracking the experience of three generations across a period of major socioeconomic transitions connected to industrialization. Like Lawrence, who was significantly influenced by evolutionary theory, Drabble examines the biological, historical, and cultural aspects of change, and from a biological perspective, the novel asks how far and how quickly human beings can adapt to changes in the environment. This question is explicitly raised by one of the three main protagonists, Bessie Bawtry, when she is at university in the 1920s and is puzzled by the frustratingly slow pace of human evolution.[8] By the 1990s, we find her granddaughter Faro writing a popular science book on 'changing concepts of evolutionary determinism', in an apparent allusion to the plethora of popular books on such topics which were published in the run-up to the HGP, for example Matt Ridley's *Genome* (1999). However, the neo-Darwinian perspective popularized by writers like Ridley is represented in the novel not by Faro but by the third-person narrator, who provides a running commentary on the characters' behaviour, frequently adverting to the genetic fate that awaits them. For example, Bessie's destiny is foretold in these terms at the very beginning of the novel:

> The Bawtrys had stuck in Hammervale for millennia, mother and daughter, through the long mitochondrial matriarchy. [...] She sensed inertia in the Bawtry marrowbone. Others had shouldered

---

[8] Margaret Drabble, *The Peppered Moth* (London: Penguin, 2001), p. 114. Further references will be given within the main text.

their pack, taken to the road, fled with dark strangers, enlisted, crossed the seas, crossed their bloodlines, died foreign deaths, spawned foreign broods. The Bawtrys had stuck here through the ages [...] And how should she, a puny sickly child, find the strength to loosen the grip of this hard land, these programmed cells? (6)

The implication is that the Bawtrys have been selected for this harsh environment and that all attempts to escape from it will fail, although the impact of this gloomy prognosis is undermined by the narrator's overblown style, which distances the reader from the claims being made and suggests that there is something absurd in this exaggerated fatalism. In this respect, Drabble may be making a point about the rhetorical flourishes that are common in books such as Ridley's *Genome*, which portentously claims, for example, that 'freedom lies in expressing your own determinism'.[9]

The novel appears at first sight to be equally sceptical about the neo-Lamarckian view of inheritance espoused by Faro. As Eve Jablonka and Marion Lamb explain, the term neo-Lamarckism was invented in 1885 but was never well defined and meant different things to different people. However, the core idea was that adaptation could occur through the inherited effects of use and disuse.[10] Many neo-Lamarckians also believed that evolution was inherently progressive and goal-directed, as did Henri Bergson, with his vitalist understanding of 'creative evolution', mentioned by Faro.[11] It is the progressive aspect of neo-Lamarckism that appeals to her, as the narrator explains:

She didn't hold with Darwinian or genetic determinism. Of course she knew that *was* how things were, but she didn't *like* the way things were. She didn't approve of it. And that was no doubt why she'd been attracted to tackle the subject in the first place. As an act of pointless but heroic resistance. A forlorn hope. She'd like to feel that one could rediscover an argument that would reinstate the freedom of the will and the adaptability of the species. (146)

---

[9] Matt Ridley, *Genome: The Autobiography of a Species in 23 Chapters* (London: Fourth Estate, 2000), p. 313.

[10] For an account of neo-Lamarckism see Eva Jablonka and Marion Lamb, *Evolution in Four Dimensions: Genetic, Epigenetic, Behavioural, and Symbolic Variation in the History of Life* (Cambridge, MA; and London, England: MIT Press, 2005), pp. 21–3.

[11] See Henri Bergson, *Creative Evolution* (1907) (London: Macmillan, 1911, 1960 printing), trans. Arthur Mitchell.

In presenting Faro's views in terms of a 'forlorn hope', the novel reflects the anti-Lamarckian prejudice which was widespread at the time when it was written. Indeed, Lamarckism is positioned as an extravagant fantasy in a discussion of the peppered moth between Faro and her boyfriend Seb, who turns out, perhaps not coincidentally, to be a liar and a fantasist. Seb takes the neo-Lamarckian position to an extreme, insisting that the moth *chose* to become darker to adapt to changing conditions: 'It grew darker', insists Seb,'It was a Lamarckian moth. It willed its own darkness. It acquired several shades of darkness. It clung on by willing its own darkness' (268–9). Faro's response is 'the man's barking mad', but she later admits to an intuition that the story of the peppered moth may have some 'hidden hope' in it, something to counteract the atomistic and determinist view of evolution which it seems to confirm.

The novel pursues the question of determinism through its detailed account of the life of Bessie Bawtry, who is born into poverty in industrial South Yorkshire in the early twentieth century, at a time when 'the pollution was so pervasive that it provoked no comment. Only strangers from the soft south or the rural northern dales noticed its pall. The natives lived in it, coughed in it, spat it out, scrubbed at it, and frequently died of it' (8). Yet she has qualities that set her apart from 'the coarser strains [that] had bred and multiplied amongst the slag heaps' as she is precocious at school and has 'needs and desires beyond her station' (7). Her intelligence and ambition mean that in theory she should be well placed to respond to the educational opportunities that were provided for 'bright' children in the inter-war period in Britain and to achieve the social mobility associated with this ladder of opportunity. However, she has enormous difficulty in making this transition, and as Drabble probes the reasons for this she uncovers the biological components of social adversity and illuminates the powerfully somatic dimensions of social disadvantage. In this respect, the novel's interests converge with those of the most recent incarnation of Lamarckism associated with epigenetics, which has shown, as noted above, that experience, particularly early life experience, can modify gene expression, and that these modifications can be inherited.

Like the child-protagonists of many *Bildungsromane*, Bessie is represented as an outsider, but what is distinctive about the novel is its focus on the somatic effects of her isolation and social anxiety. As a child she feels at odds with her family and with the grimy surroundings of Breaseborough and aided by a 'delicate constitution' she falls into the habit of illness, escaping to the 'dark cave' of her bed. This is a pattern which is repeated at a crucial period in her adolescence when she is interviewed for a place at Cambridge. She is haunted by this experience, which is described in terms suggestive of the flaying of human skin: she feels 'humiliation, grief, rawness, her very skin aflame with tenderness, her clothes exposed, her accent exposed, amidst all those confident southern girls from boarding schools'. The emphasis on somatic pain is continued in the description of the Principal's 'insulting condescension', as her 'tone burns'. Bessie is mortified to discover that in the Principal's eyes her County Scholarship is a mark of shame rather than triumph as she sees it 'held up for inspection as though it were a damp kitchen rag, a servant's dishcloth instead of a laurel wreath' (101). A little later, she is invited to a house party in Breaseborough by a local patron of the arts, an experience she finds even more traumatic. Again we learn of her feelings of humiliation and shame because she is unable to dance or hold her own in conversation; again the pain is written on her body, on her 'smarting eyes, her stinging red nose' (101). Immediately afterwards she collapses, feeling that she is an 'unutterable failure' (91). At this point the narrator describes Bessie as a pupa, underlining the fact that this is a critical and potentially transformative stage in her development:

> She lay still, in turmoil. A seething, a pregnant brewing, a splitting, a proliferation of particles. Is it a sickness, is it a fermentation, is it a couching, and what will it bring forth? Is it a growing or a dying? (98)

Amidst this imagery of birth and death, the neo-Darwinian idea of the emergence of a favourable mutation is briefly introduced, in a fantasy of (genetic) adaptation during the life-course of an individual. However, it is immediately abandoned as the narrator states flatly that 'Bessie did not mutate'. Moreover, there is a suggestion of long-term hidden damage which is intensified in the account of a second collapse when Bessie is at university. She shows 'vague signs of distress...tell-tale signals that anyone familiar with her medical track record would have noted with alarm' (115–16) and once again takes to her bed. The immediate trigger is that she has prepared for the wrong exam, but the narrator stresses that this has happened because she is 'all at sea' at Cambridge, lacking the social and cultural capital needed to navigate the university system: she feels like a 'slack, stupid, dumb girl from the north' (121).

As an adult, Bessie is diagnosed with 'endogenous' depression, a now-obsolete term for a form of depression which was not thought to be linked to any obvious external factors, but as the novel shows, her depression is in fact inseparable from such external forces. Repeatedly, her experience of psychosocial stress is followed by a somatic collapse which in turn morphs into depression, suggesting that her illness is not the result of genetic predetermination but may be an epigenetically mediated response to stress. The fact that it is particularly acute in adolescence and early adulthood matches the concept of developmental plasticity which is associated with epigenetics and which emphasizes that in humans, plasticity extends beyond infancy into adolescence and early adulthood.[12] More specifically, Bessie's experiences can be linked to the concept of 'social defeat', which first emerged in animal studies but which has subsequently been extended to research in humans, where it has been found that repeated exposure to social aggression or humiliation is associated with depression and generalized anxiety disorder. Epigenetic modifications are emerging as the primary mechanism through which such experiences are mediated, as gene expression is altered to produce responses which may or may not be adaptive.[13] In Bessie's case, her early depression could be considered adaptive, as her illness secures her time and space to recover and sit her exams. In her later life, however, it recurs with increasing severity, particularly at times of social pressure, as when she moves to an affluent part of Surrey where she is surrounded by neighbours whom she fears

---

[12] The classic work on developmental plasticity is Mary-Jane West Eberhard's *Developmental Plasticity and Evolution* (New York: Oxford University Press, 2003). Research in epigenetics adds further support to her contention that plasticity allows organisms to adapt to the environment and may also be a driver of evolution.

[13] See Frances A. Champagne, 'Epigenetic influence of social experiences across the lifespan', *Developmental Psychobiology* 52, 4, May 2010, 299–311. doi: 10.1002/dev.20436. See also McGowan et al., 'Epigenetic regulation of the glucocorticoid receptor in human brain associates with childhood abuse', *Nature Neuroscience*, 12 (3) (2009), 342–8. doi:10.1038/nn.2270

and dislikes. At this point she sinks into depression 'with an almost voluptuous abandon', subsequently becoming agoraphobic.

As the novel reflects on the legacy of Bessie's distress for her daughter and granddaughter, it broaches the issue of transgenerational epigenetic inheritance. There is an increasing body of evidence to show that epigenetic modifications to the genome can be transmitted between generations. For example, Michael Meaney's much-cited work has shown that maternal licking of rat pups causes epigenetically-mediated changes in the developing brain, creating calmer, less fearful adult females that are more likely to lick their own pups.[14] Such changes play a crucial role in the developmental process and offer alternative paths of inheritance which confirm the neo-Lamarckian view that heritable adaptive change can occur in response to the conditions of life. Drabble's text is particularly interested in such non-genetic inheritance, which is explicitly linked to Faro's neo-Lamarckian interests. There are frequent references to Lamarck in the sections of the novel devoted to Bessie's daughter Chrissie and to Faro, and in line with this perspective there are repeated suggestions that life experience may in some sense be 'remembered' by the body and passed on to future generations. There is a suggestion that Chrissie has inherited her mother's response to social defeat when the rhetoric of being flayed and publicly humiliated, previously associated with Bessie's experiences at Cambridge, recurs in the description of Chrissie's desolation when her first husband leaves her. Elsewhere there is explicit speculation about forms of non-genetic inheritance. For example, when Chrissie resists the claims of a flamboyant and seductive boyfriend, the narrator suggests that 'some remnant of her sensible, Yorkshire, Bawtry-Barron self' may have clung to the belief that she ought to get her degree and asks, rhetorically, 'had something in her remembered Joe Barron's two years as a travelling salesman, and Bessie's collapse before her Part One English B?' (246). When Faro is mulling over how and why some members of her family came to escape from Breaseborough, the text invokes the possibility that experience may be transmitted across three generations. Faro concludes that this escape required 'a movement of the mind' and at this very moment, according to the

---

[14] Weaver et al., 'Epigenetic Programming by Maternal Behaviour', *Nature Neuroscience* 7 (2004), 847–54.

narrator, 'a memory, just beyond retrieval, like a shadow of an unremembered dream, is nagging at Faro. But it is not her memory. The memory is not hers, so she cannot remember it. It is not in reach. It hovers and flickers, with a faint colouring, a rustling, an inarticulate appeal' (147). Through the reference to 'a movement of the mind' the text suggests that this 'shadow of an unremembered dream' is linked to Bessie's attempts to escape from Breaseborough decades earlier.

The Peppered Moth is not, in fact, quite a novel. Its epigraph is a poem by Drabble's daughter Rebecca Swift focusing on her memories of her maternal grandmother, and it is followed by an afterword in which Drabble describes it as 'a novel about my mother' (390). It could be seen as a form of refracted life-writing, as Drabble reflects not only on her mother's depression but on her own, a theme which is explored in more detail in *The Pattern in the Carpet*, which combines a history of the jigsaw with reflections on the way in which Drabble has at various points in her life used jigsaws to combat her own depression. In *The Pattern in the Carpet* she comments on the way in which her mother's depression 'infected' her and notes that *The Peppered Moth* represents her attempt to understand this infection in terms of genetic inheritance.[15] However, as the novel reflects on Bessie's history and locates it in the context of broader social and economic transitions, the plot of genetic inheritance is thickened by the evocation of the somatic imprint of early experience. Anticipating the science of epigenetics in this respect, the novel also anticipates some of the major questions epigenetics raises in relation to social equality. When epigenetics first appeared on the horizon, it was welcomed by many in the medical and social sciences because by stressing the role of the environment in shaping the phenotype, it underscored the way in which social and political factors shape health inequalities: in contrast with the genetic determinants of ill-health, socio-political factors are, in principle, open to progressive intervention. However, as the anthropologist Margaret Lock has argued, in practice epigenetics research has entailed a miniaturization and a molecularization of 'the environment'. Drawing on the work of Jorg Niewöhner, she contends that in the research laboratory, in order to establish correlations between epigenetic changes and

---

[15] Margaret Drabble, *The Pattern in the Carpet: A Personal History with Jigsaws* (London: Atlantic Books, 2009), pp. 26, xiii.

the social environment, events of significance in people's lives are standardized so that they can be assimilated to a quasi-natural experimental system. In the process, as she puts it, 'subjective knowledge and meanings attributed to events are ultimately ignored, washed out, and rendered insignificant'.[16] In addition to this flattening out of social contexts, there is the danger that genetic determinism might be replaced by epigenetic determinism, together with a neo-Lamarckian view of some populations as being permanently damaged by environmental toxins, including stress, thus constituting a new social underclass. This raises the question of the wider implications of epigenetics for our understanding of social justice. As Maurizio Meloni has pointed out, many theories of justice, most notably that of John Rawls, depend on the distinction between the natural and the social, whereby what is 'natural' is deemed to lie beyond the remit of social responsibility. However, by providing the missing link between the natural and the social, epigenetics breaks down the distinction between these domains. If a given condition is caused by an epigenetic response to environmental toxins, it can no longer be considered simply natural but must also be viewed as profoundly social, which raises the question of whether society has a responsibility to rectify or compensate for such a condition.[17] In this context, where do the boundaries between the individual and the collective lie? In addition, what are the implications for transgenerational justice? As Drabble shows in *The Peppered Moth*, the Industrial Revolution created an environment in which for the vast majority, socioeconomic deprivation was accompanied by pollution and poor living conditions. In Bessie's case, deprivation translates into social anxiety and depression, but Drabble also draws attention to the effects of pollution through the case of her husband Joe, who contracts lung disease as a result of industrial pollutants: crucially, both conditions are linked to epigenetic effects which may be inherited across generations. If the social and material contexts of past generations are somatically transmitted in this way, perpetuating

---

[16] Margaret Lock, 'Comprehending the Body in the Era of the Epigenome', *Current Anthropology* 56, April 2015, 151–77 (160).
[17] Maurizio Meloni, 'Epigenetics for the social sciences: justice, embodiment, and inheritance in the postgenomic age', *New Genetics and Society* 34, 2015, 125–51 (132). doi:10.1080/14636778.2015.1034850.

disadvantage, might succeeding generations be entitled to some form of reparative justice? In tracing intersecting forms of biosocial inheritance, the novel raises questions about the potential need to broaden the remit of social justice to include natural justice.

Jackie Kay's much-admired memoir *Red Dust Road* broaches similar themes, which are explored through the crucible of her experience of transcultural adoption.[18] Kay has a Scottish mother and a Nigerian father and was subsequently adopted by white working-class Scottish parents. Like many adoptees, she began to search for her birth parents when she was herself about to have a child and *Red Dust Road* is a retrospective account of this quest for origins. As Margaret Homans has pointed out, the experience of adoption confronts us with the question of the unknowability of origins in a particularly acute form.[19] Moreover, because Western cultures equate genetic origins with identity, adoption is commonly viewed as a profoundly traumatic rupture. In a gesture which is common in adoption narratives, Kay's memoir both rehearses and challenges the genetic model of inheritance and the beliefs about identity that are associated with it. Her search for her birth parents is initially articulated in the language of genes and blood but as the text unfolds it runs up against the limits of the explanatory power of genetics and begins to shape an ontology which resonates powerfully with the insights of epigenetic science. As we have seen, research in epigenetics suggests that identity is not the result of a fixed programme but is shifting, dynamic, and produced through continuous interactions between the individual and the environment. In biological terms, it is construed in terms of plasticity, a concept which has significant philosophical implications. Catherine Malabou has argued that neuronal plasticity, in particular, generates a view of the individual as a 'self-sculpting' organism and has mapped the congruencies between the plasticity of the brain and 'the possibility of

---

[18] The term transcultural is used here in preference to transracial as a more flexible term which can accommodate the complexity of the racial, national and sociocultural identities which the adoptee encounters, especially in the context of the migratory conditions of recent history. See John McLeod, *Lifelines: Writing Transcultural Adoption* (London: Bloomsbury, 2015) for a thoughtful discussion of this issue (pp. 7–10).

[19] Margaret Homans, 'Adoption Narratives, Trauma, and Origins', *Narrative* 14 (January 2006), 4–26 (5).

fashioning by memory [...] the capacity to shape a history'.[20] This account reads the history shaped in Kay's memoir in relation to the biological and philosophical meanings of plasticity and argues that in her understanding of the mutable and entangled nature of being, she forges a postgenomic ontology.

When Kay first meets her birth parents the question of physical resemblance grounded in shared DNA is her most immediate concern. When she meets her mother, she is touched by the fact that they both bring orchids to the meeting and wonders if that fact has anything to do with genetics, but at the same time is taken aback by the lack of any physical similarity between them. This lack or absence creates a profound sense of disorientation. No matter how hard she 'stares' at her mother she can't see 'anything of my face in hers' and in consequence the meeting takes on a fake quality: she writes that 'there's an uncomfortable intimacy between us; the knowledge that we are biologically mother and daughter makes me feel a little bit like a fraud'.[21] Her meeting with her birth father is equally disorientating, despite or because in him she finds the physical resemblances for which she was searching. These similarities, paradoxically, are equally disorientating; her father's physical presence seems to cancel her out, rendering her insubstantial and ghost-like. She comments that the 'ordinary fact [of resemblance] that most people take for granted is spooky and strange for an adopted person' (135). Her choice of words suggests that for an adopted child, physical resemblances are uncanny in the Freudian sense, undermining their sense of possessing a unique identity.[22] In this text encounters with birth parents thus work to unravel rather than endorse the mythology of genetic kinship. This is predicated on the

---

[20] Catherine Malabou, *What Should We Do With Our Brains?*, foreword Marc Jeannerod, tran. Sebastian Rand (New York: Fordham University Press, 2008), p. 6. Further references will be given within the main text.

[21] Jackie Kay, *Red Dust Road* (London: Picador, 2010), 65. Further references will be given within the main text.

[22] However, as McLeod notes, Kay also records the very different experience she has when she meets her half-brother Sidney in Nigeria: she writes 'I feel a strange almost ecstatic sensation of recognition. It is nearly primitive. I could happily sniff his ears and lick his forehead. It has completely ambushed me; I wasn't expecting it at all' (p. 272). Kay is open to the experience of somatic recognition reported by many adoptees, which is enabled here by Sidney's warm and immediate acceptance of her as his sister, but she does not endow these feelings with any overarching or ultimate significance.

belief that individuals who are related by blood have will similar traits, which will in turn generate 'natural' or instinctive affinities, but Kay experiences no such feelings of recognition, writing that:

> The whole business of being adopted seems on the one level to be a fantastic fiction. Something about it, even to you, seems fake. Anybody could be sitting there saying they are your mother or father; you could be anybody claiming to be his or her daughter. Yes, on some level it seems fabulously made up, as attractive as an idea, as it is revolting, that this total stranger sitting next to you, should be related to you by blood. (134–5)

Kay here problematizes the concept of blood relationship and at the same time points to new ways of thinking about biological ties. The idea that a total stranger could be recognized as an intimate relation is an affront to experiential knowledge (the idea is 'revolting'); folded into this insight is the intuition that biological connections do not pre-exist experience (as genes were once thought to do) but are forged in and through the processes of life.

As she explores these processes, Kay draws out the entanglement of the somatic and the social in ways which echo Lock's concept of 'local biology' first articulated in the 1990s to capture the way in which the embodied experience of sensation is in part informed by the material body, which is in turn shaped by evolutionary, environmental, social, and individual factors. Crucially in this model, embodiment is also shaped by the way in which self and others represent the body, drawing on local categories of knowledge and experience.[23] The idea of 'local biology' is especially helpful for reading Kay's memoir because of its emphasis on the inextricability of the biological and the symbolic in the

---

[23] Lock first developed the concept of local biology in *Encounters with Aging: Mythologies of Aging in Japan and North America* (Berkeley: University of California Press, 1993). She has been at pains to emphasize in recent work that the concept of local biology does not imply any fixed link between individual populations and the environment, rather, the body and the environment co-construct each other across time. The point is an important one because of the danger of stigmatizing particular populations or reifying environmental damage, noted by Meloni in 'Epigenetics for the social sciences'. For a discussion of the fluidity of local biology, see Margaret Lock, 'Recovering the Body', *Annual Review of Anthropology* 46, 1–14, October 2017. doi: 10.1146/annurev-anthro-102116-041253, https://www.annualreviews.org/doi/full/10.1146/annurev-anthro-102116-041253 [accessed 13 May 2019].

formation of the self, a point which Kay emphasizes in her account of her early childhood in *Red Dust Road*. The first story she has of her own origins comes from her adoptive mother Helen, who describes the process in terms that mirror the stories of 'kinning' analysed by Signe Howell in her classic study of transnational adoption in Norway.[24] Howell suggests that the process of adoption involves the construction of a symbolic pregnancy and birth. Pregnancy begins when a couple first decides to register with an adoption agency and it involves creating 'emotional space' for the anticipated child. Birthing starts when the child has been allocated, alongside the 'kinning' process which folds them into the adoptive family as photographs and other details are circulated among family and friends. Something similar happens with Helen Kay, who has what Kay describes as a 'ghost pregnancy' where she shadows Kay's birth mother in her imagination. Kay writes that:

> months before my birth mother gave birth to me, my mum knew that she was going to have me. 'It was the closest I could get to giving birth myself,' she's told me often. 'I didn't know if I'd have a girl or a boy, if you'd be healthy or not, the kind of thing that no mother knows. It was a real experience. It felt real. I remember waiting and waiting for news of your birth and phoning up every day to find out if you'd been born yet.' (26)

When she is born, however, Kay is not healthy: she has been injured by a forceps delivery, is thought to have brain damage and is not expected to live. Her adoptive mother is told to pick another baby but refuses because she has become attached to the idea of this baby, 'her' baby. So far in Kay's story (which is, of course, the story of her adoptive mother's story) the connection between mother and daughter is framed as a purely social relationship, but from this point on a link is posited between biology and culture, as Kay speculates that the affective bond between them is responsible for the fact that she thrives, unexpectedly, in the first months of her life:

> [My mother] visited every week, or every month, depending when she's telling the story, driving the forty miles from Glasgow to

---

[24] Signe Howell, *The Kinning of Foreigners: Transnational Adoption in a Global Perspective* (Oxford: Berghahn Books, 2007). Further references will be given within the main text.

Edinburgh, with my dad, and she had to wear a mask, so as not to infect me, and got to pick me up and hold me. Perhaps this interest, this love, is what made me survive against the odds. The doctors were apparently amazed at my recovery. (48)

So whereas Howell found that adoptive families start from a focus upon biology as the model for the family but end up by asserting a social-constructionist approach to family-formation, Kay's story has a different narrative arc (70). Rather than privileging the socio-cultural over the biological aspects of kinship, the palimpsestic narrative shaped by Helen and Jackie Kay is sensitive to the multiple factors that shape identity and to the inter-relations between social meanings and somatic experiences (the very fact that it's a *story* of embodied care rather makes the point). Kay's story of origin can thus be read as a fable which dramatizes the way in which nurture becomes embodied; moreover, as a narrative which illuminates the embodied connection forged through an adoptive mother's nurture of her child, it complicates a narrowly gene-centric view of biological relatedness.

Her parents' loving support enables Kay to flourish, but the story of her brother's early life broaches more painful issues relating to the somatic impact of racism and discrimination. As mentioned in the previous chapter, in the wake of the completion of the Human Genome Project in 2003, it was widely assumed that the concept of the genetic basis of race had been shown to be false.[25] Nonetheless, 2004 saw the first licensing of a drug targeted at a specific racial group, African Americans. This was BiDil, a combination of two existing drugs which was authorized to treat heart failure in self-identified black patients. As Dorothy Roberts has eloquently argued, such race-based medicine re-inscribes a biological understanding of race and opens the door to a new biopolitics of race.[26] However, in the context of the debates about race-based medicine, Shannon Sullivan underscores the

[25] Following the completion of a draft map of the human genome in June 2000, Craig Venter (Head of Celera Genomics and the chief private scientist involved with the HGP) claimed that his analysis of the genomes of five people of different ethnicities had demonstrated that 'race' was not a scientifically valid construct. See C.J. Venter (26 June, 2000) Remarks at the Human Genome Announcement, https://clintonwhitehouse5.archives.gov/WH/New/html/genome-20000626.html [accessed 29 April 2019].

[26] Dorothy E. Roberts, 'What's Wrong with Race-Based Medicine?: Genes, Drugs, and Health Disparities', *Minnesota Journal of Law, Science and Technology* 12, 2011, 1–21 (10).

importance of epigenetics in providing a more nuanced understanding of the impact of social forces such as racism on health. As she argues, the so-called *racial* health disparities which race-based medicines purport to address are more accurately called *racist* health disparities, as they cannot be explained by race apart from the phenomena of discrimination and prejudice. Developing this point, she draws on the work of Chris Kuzawa and Elizabeth Sweet which suggests that when environments are differentially constructed according to socially imposed racial identities, patterns of health disparity are created through epigenetic mechanisms.[27] Specifically, they argue that prenatal epigenetic changes are responsible for the low birth weight which is disproportionately common among African Americans and which is a predictor of subsequent cardiovascular disease. Such epigenetic changes are prompted by social factors, especially those relating to the experience of racism and inequality, and like many epigenetic modifications they appear to be heritable, meaning that psychosocial stress can have a direct impact on the physical and mental health of the next generation. As Sullivan acknowledges, one of the dangers of such research is that it could be interpreted as 'proof' of the diseased or broken black family, but equally, she argues that epigenetics can be recruited as part of a 'critical arsenal' for combating racism by fully comprehending its extent. Epigenetics has revealed that the damage caused by racism is more extensive than was previously thought, as the impact of psycho-social stress is literally embodied; the further implication is that the fight against racism must be waged 'on biological and medical levels, as well as economic, political, aesthetic, and other social terrain'.[28]

Kay's memoir takes up these issues in its reflections on her brother's adoption story. The narrative Kay's adoptive mother creates around Maxie's origins is vivid and disturbing, as she tells of 'black men coming off boats, stories of black blood' and explains that when he

---

[27] Chris Kuzawa and Elizabeth Sweet, 'Epigenetics and the Embodiment of Race: Developmental Origins of US Racial Disparities in Cardiovascular Health', *American Journal of Human Biology* 21 (2009), 1–15. doi: 10.1002/ajhb.20822. See also Grazyna Jasienska, 'Low birthweight of contemporary African Americans: An intergenerational effect of slavery?', *Am. J. Hum. Biol.* 16 (2009), 16–24. doi:10.1002/ajhb.20824.

[28] Shannon Sullivan, *The Physiology of Sexist and Racist Oppression* (Studies in Feminist Philosophy) (New York: Oxford University Press, 2015), p. 127.

was adopted, tests were done on him in Edinburgh to see if he had any 'negro blood' because they 'couldn't explain his colour' and it was finally decided that he was what was known as a 'throwback' (such tests being a tangible marker of biomedical racism). Kay speculates that it was the fantasy of 'negro blood' which led his birth parents to place him for adoption, a suspicion which is supported by the fact that the couple go on to marry after they have placed Maxie in an orphanage. Kay suggests that this foundational experience of rejection became an integral part of her brother's identity:

> I often wonder if the story of the hospital having tests done on his blood and then his parents putting him into an orphanage is a story that he carried with him, long before he ever heard it in words, if it is possible that we imbue ourselves and our personalities with our own stories, pre-articulation, so the story turns into a kind of inheritance.
> (198–9)

In this subtle formulation, which hesitates between the registers of the biological and the social, Kay captures the way in which discrimination can become internalized, can become part of an embodied narrative long before the acquisition of language. Exploring the permeable boundaries between psyche and soma, she maps the way in which rejection in early life predisposes Maxie to the feeling that he is being 'persecuted', creating a psycho-biological feedback loop which is exacerbated by his subsequent experiences of racism. When he is a teenager his albino pet rabbit (which is like him an outlier in genetic terms) is mysteriously strangled and a few years later an anonymous stranger pours petrol on his motorbike and sets it on fire. Kay notes that 'things like that happened to him throughout his life, a ghastly cocktail of bad luck and racism' (200). The 'cocktail' generates an all-pervasive feeling of insecurity so that when he goes for job interviews, he is sure that he is being discriminated against because his adoptive father is a communist. As Kay comments, 'it was all too much for him. He didn't want anything that made him stand out more than he already did' (199).

Building on these insights into the complexity of development, the text goes on to articulate an ontology which resonates with Malabou's account of the mutability of being in *What should We Do with Our Brain*. Drawing on the work of the neuroscientist Jean-Pierre

Changeux, Malabou argues that neuronal plasticity, in which networks are formed and re-formed in response to experience, supports an understanding of subjectivity in terms of self-making rather than genetic determination.[29] Plasticity is identified with 'suppleness, a faculty for adaptation, the ability to evolve', and is contrasted with flexibility, which signals the reception but not the giving of form and hence a kind of 'docility', particularly in the context of a global economy which demands adaptability without limits (12–13). Plasticity demonstrates the brain's capacity to make its history and become the subject of its history and according to Malabou, conscious awareness of this 'freedom of the brain' can extend the potential for individual and collective self-transformation, for 'grasping the connection between the work of genetic nondeterminism at work in the constitution of the brain and the possibility of a social and political nondeterminism, in a word, a new freedom' (13). The counterpart of such positive plasticity is the destructive plasticity which she discusses in *Ontology of the Accident* and *The New Wounded*, where she explores forms of neuronal damage which create an absolute break in personality, rendering us strangers to ourselves.[30] However, in *What Should We Do with Our Brain?*, she turns to Boris Cyrulnik's concept of resilience to draw out the regenerative potential of plasticity. For Malabou, resilience is 'a logic of self-formation starting from the annihilation of form', a process of self-reconfiguration which is developed 'simultaneously against and with the threat of destruction' (76). She cites as an example work with orphans in the institutions of the Ceauçescu regime in Romania, which showed that early lack of care left epigenetic traces which were associated with impaired mental and physical development. However, if the children were fostered soon enough, many of the changes were reversed, signalling the fact that 'cerebral traces are reparable'.[31]

---

[29] Jean-Pierre Changeux, *Neuronal Man: The Biology of Mind*, new edn, trans. Laurence Galey (Princeton: Princeton University Press, 1997).

[30] See Catherine Malabou, *Ontology of the Accident: An Essay on Destructive Plasticity*, trans. Carolyn Shread (Cambridge: Polity, 2012) and *The New Wounded: From Neurosis to Brain Damage*, trans. Steven Miller (New York; Fordham University Press, 2012).

[31] Boris Cyrulnik, quoted in Malabou, *What Should We Do with Our Brain?*, p. 77. For an account of a large-scale study of the effects of neglect in Romanian orphanages see Charles A. Nelson, Nathan A. Fox, and Charles H. Zeanah, *Romania's Abandoned Children: Deprivation, Brain Development, and the Struggle for Recovery* (Cambridge, MA: Harvard University Press, 2013).

Red Dust Road proposes a similarly dynamic ontology, which comes to the fore in Kay's account of an accident which she has at the age of sixteen, at the point when she is revising for her O-Levels. In line with one of the text's governing metaphors, the roads which proliferate and branch off through life, Kay stresses the adventitious nature of the decision which changes her life. Driving her Honda moped, she takes a turn she has never taken before and is knocked off her bike, breaking her leg in two places: she is hospitalized and operated on, but it is eighteen months before she walks properly again. As she explains, extending the metaphor of the road to that of the border, the accident makes her think a lot about death: 'getting close to death, it seems to whisper at the edge of your cheek. Nearly dying brings you closer to living. There's a thin border: you feel yourself cross it, going back to the land of the living, going home' (233). In addition, the accident alters her metabolism, so that she is no longer a slim athlete who trains for the Scottish County Championships; instead she puts on weight and acquires a permanent limp. The accident is traumatic in psychological terms, separating her from her previous self but as Kay notes, echoing Malabou's point about reconfiguring the self 'against and with the threat of destruction', 'accidents, if you don't go and bloody die, can offer up a new way to live' (236). In her case, it is the broken leg and the long period of convalescence that make her into a writer: as she explains, 'I suddenly saw the world differently and I knew that I wanted to write about what I saw.' She reads voraciously and writes poems about 'the accident, and apartheid and poverty and peace and housewives and anything else that interested me'; her English teacher then shows them to the writer and artist Alasdair Gray, who tells her there's no doubt in his mind that she is a writer. Kay represents her metamorphosis in terms that resonate both with Malabou's emphasis on the regenerative potential of neuronal networks and Cyrulnik's definition of resilience as a mesh, not a substance: as he puts it, 'we are forced to knit ourselves, using the people and things we meet in our emotional and social environments'.[32]

The imbrication of writing and self-sculpting is especially marked in Kay's text and resonates with Malabou's analysis of the connections

---

[32] Boris Cyrulnik, *Resilience: How Your Inner Strength Can Set You Free from The Past*, trans. David Macey (London: Penguin, 2009), p. 51.

between neuronal plasticity and writing. A former student of Derrida, Malabou deploys plasticity in *Plasticity at the Dusk of Writing* as a means of critical engagement with deconstruction and with the concept of writing as *différance*. She points out that the deconstructive project was the result, in part, of developments in linguistics, genetics, and computer science and suggests that the genetic code 'became a true ontological motif' when the geneticist Francois Jacob claimed that any material structure could be compared to a message: this confirmed an understanding of the organization of the real in terms of 'a global morphology made up of "meaningful gaps" and differences'.[33] However, the power of this linguistic scheme is diminishing as plasticity is 'slowly but surely establishing itself as the paradigmatic figure of organisation in general' (59). Referring to Changeux's *Neuronal Man*, Malabou notes that neuronal organization is now conceptualized in terms of assemblies or cooperative groups of neurons with the potential to recombine amongst themselves. The implication of this is that though synaptic fissures are certainly gaps, they are '*gaps that are able to take form or shape*'. She continues 'that's it, in fact: *traces take form*. It is striking to note that neuronal plasticity—in other words, the ability of synapses to modify their effectiveness as a result of experience—is a part of genetic indetermination. We can therefore make the claim that *plasticity forms where DNA no longer writes*' (60). Throughout her work Malabou holds to a conception of plasticity as the originary mutability of being and to the contention that 'to behold essence is to behold change'.[34] Kay's text embodies such plasticity, as it replaces a linear narrative with a dynamic mapping of points of contact between the

---

[33] Malabou, *Plasticity at the Dusk of Writing*, p. 58. Further references will be given within the main text. See also Jacques Derrida's comments on the relationship between deconstruction and molecular biology in the introduction to *Of Grammatology*, trans. Gayatri Chakravorty Spivak (Baltimore and London: Johns Hopkins University Press, 1976). In his discussion of what Malabou calls the '*semantic enlargement* of the concept of writing' he notes that 'it is also in this sense that the contemporary biologist speaks of writing and *pro-gram* in relation to the most elementary processes of information within the living cell' (p. 9, italics in originals).

[34] Catherine Malabou, *The Heidegger Change: On the Fantastic in Philosophy*, trans. Peter Skafish (Albany, NY: SUNY Press, 2011), p. 16, quoted in Brenna Bhandar and Jonathan Goldberg-Hiller, eds, *Plastic Materialities: Politics, Legality and Metamorphosis in the Work of Catherine Malabou* (Durham and London: Duke University Press, 2015), p. 4.

symbolic and the material and between past, present, and future. It shapes a plastic temporality, in which the present time of the narrative is invaded by the past, which is in turn modified by the present, as both inflect the imaginary future. It also captures something which as Ian James has argued is even more fundamental to Malabou's thought than plasticity, which is the void or ontological groundlessness.[35] The play of being in *Red Dust Road* rests on an apprehension of the abyss of being which as Kay suggests is intensified by the experience of adoption: it is figured in the trope of the 'wild and windy place' she imagines as preceding her adoption, which is 'like Wuthering Heights, rugged and wild and free and lonely. The wind rages and batters at the trees' (45).

Research in epigenetics also demands a radical reconsideration of the neo-Darwinian view that evolutionary change depends on random genetic mutation and involves competitive struggle between 'selfish' genes. The accumulating evidence for environmentally triggered modifications to the genome suggests that evolutionary change may be driven by phenotypic responses to the environment in addition to selection among genes.[36] Moreover, epigenetic insights into the complexity of development are generating a holistic view of life as evolving in nested systems of cells located within organisms within wider ecosystems. As John Dupré suggests, such systems can be viewed as cooperative rather than competitive as the elements in a cell or the cells in a multicellular organism 'work in a highly coordinated way and subordinate their own "interests" to those of the whole of which they are a part'.[37] In addition, cooperation between organisms is evident in numerous instances of symbiosis, most notably as mitochondria were formed by the fusion of bacteria with eukaryotic cells (endosymbiosis). It is also evident that some of the so-called 'junk' DNA which was inserted into the human genome by viruses can play a crucial part in regulating gene expression.[38] Such findings raise questions about the

---

[35] For a discussion of Malabou and the void see Ian James, '(N)europlasticity, Epigenesis and the Void', *parrhesia* 25, 2016, 1–19 (6).

[36] For a discussion of some of the implications of epigenetics for evolutionary theory see John Dupré, *Processes of Life* (Oxford: Oxford University Press, 2012), pp. 235–6.

[37] Dupré, *Processes of Life*, p. 150.

[38] The ENCODE project was set up in 2003 to address the question of the functionality of the genome, including 'junk DNA', that is, the more than 98 per cent of DNA which does not code for protein. The initial results were announced in 2021; see the ENCODE

boundaries between supposedly distinct organisms and require us to re-orient our view of the relationship between human beings and other organisms, seeing them/us not as free-standing entities but as caught up in complex relationships of inter- dependency.

The motif of cooperation is brought into Kay's memoir by virtue of her birth father's profession as an ethnobotanist who works on the mutual relationship between human culture and *Moringa oleifera*, a tree which is common in sub-Saharan Africa and which is used in traditional medicine. As Kay learns more about her father's early life and about ethnobotany more generally, she discovers extensive evidence of cooperation within and between species, noting for example that:

> Trees are so benevolent. I've just finished learning about trees working in the Forest of Burnley, how trees compensate for each other; if one is growing a little to the east, the other will move a little to the west to make room. How trees breathe the same air and are aware of each other's company. How they complement each other's growth, how two ash trees might share a canopy of leaves. (289)

Such insights are in line with current research on the cooperative distribution of resources in trees and they also resonate with thinking about inter- and intra- species cooperation which can be traced back to the Russian zoologist Karl Kessler's 1880 lecture 'On the Law of Mutual Aid'.[39] Following Kessler, there is a long history in Russian biology of

---

Project Consortium, 'An integrated encyclopedia of DNA elements in the human genome', *Nature* 489, 7414 (2012), 57–74. The consortium showed that the human genome is pervasively transcribed even when noncoding, and that the transcripts are involved in many forms and levels of genetic regulation that had not previously been suspected. In consequence, Evelyn Fox Keller argues, we have moved from 'an earlier conception of the genome (the pregenomic genome) as an effectively static collection of active genes (separated by "junk" DNA) to that of a dynamic and reactive system (the postgenomic genome) dedicated to the regulation of protein-coding sequences of DNA'. Evelyn Fox Keller, 'The Postgenomic Genome', in *Postgenomics: Perspectives on Biology and the Genome*, eds Sarah Richardson and Hallam Stevens (Durham and London: Duke University Press, 2015), pp. 9–31 (10). Some geneticists have contested these findings: see W. Ford Doolittle, 'Is Junk DNA Bunk? A Critique of ENCODE,' *Proceedings of the National Academy of Sciences* 110 (2013), 5294–300. doi:10.1073/pnas.1221376110.

[39] See for example Bingham et al., 'Ectomycorrhizal Networks of Pseudotsuga menziesii var.glauca Trees Facilitate Establishment of Conspecific Seedlings Under Drought',

interpreting biological systems in terms of mutuality rather than competition, a tendency which Daniel Todes has linked with the distinctive social and geographical landscape of Russia, 'a vast, under-populated continental plain [...] in which only the struggle of organisms against a harsh environment was dramatic'. In contrast, he points out, Darwin's theory of evolution originated in a crowded island with a capitalist economy and individualistic culture, so that the 'struggle for existence did not seem a metaphor at all, but, rather, a simple and eloquent description of nature and society'.[40] Todes highlights the complex cultural contexts of Darwinian and neo-Darwinian theory and also points to the integral role of metaphor, which melds science and life experience, in animating scientific research. In this respect, it is telling that the Darwinian metaphor of the tree of life, the traditional way of representing the relationships between organisms, is currently giving way to the dynamic, multidimensional metaphor of the web.

As Kay's memoir unfolds, her experience of adoption is grounded in this expanded biological perspective. The emphasis shifts away from the genetic origins which are represented by the trees at Quarry Mill Bank which Kay visits before and after meeting her father. On her first visit, she endows a beech tree with fairy tale properties as she slips a coin inside it as an offering before leaving for Nigeria, but on the second, she slips a coin inside an oak tree which is shedding its leaves, signalling loss. A fantasy of origin which is linked both to the family tree and to the Darwinian tree of life has been lost but is replaced by an alternative understanding of growth and development which is represented through the metaphor of transplantation. When Kay's adoptive mother first brings her brother home from the orphanage, she plants a cherry tree to celebrate his arrival and fifty years later Kay plants the

Ecosystems 15 (2012), 188–99, doi: 10.1007/s10021-011-9502-2 at http://link.springer. com/article/10.1007%2Fs10021-011-9502-2 [accessed 15 January 2014]. See also Peter Wohlleben's popular book *The Hidden Life of Trees: What They Feel, How They Communicate: Discoveries from a Secret World*, trans. Jane Billinghurst (2015) (London: William Collins, 2017). Wohlleben argues that trees equalize their resources via underground root networks (p. 16).

[40] Daniel Todes, 'Global Darwin: Contempt for Competition', *Nature* 462, 7269 (2009), 36. doi: 10.1038/462036a. For further discussion of the social contexts of twentieth-century genetics see Clare Hanson, *Eugenics, Literature and Culture in Postwar Britain* (London: Routledge, 2012), chapter 3.

seeds of *a Moringa oleifera*, given to her by her Uncle Nwora in Nigeria, to celebrate finding her brother Sidney. She writes:

> I try and imagine my own moringa growing in the front garden of my terraced house in Manchester, far away from its home. I wonder if the pods will ever take root. I picture my healthy, tall moringa in ten years' time. Will I still be living here then? I imagine a magical moringa, years and years away from now; its roots have happily absorbed and transported water and minerals from the dark, moist soil to the rest of the splendid tree. (223)

Both these trees have their origins in far off places: the moringa comes from Africa and South Asia while the cherry comes from East Asia, but they thrive when they are transplanted into rich new soil. Adoption is folded into an understanding of flourishing which does not depend on genetic origin but on an organic process of accommodation to change.

In this respect, Kay's memoir also emphasizes the porosity of the body and its openness to the environment. She seeks the traces of her 'ancestral environments', visiting the house where her mother grew up and finding the landscape of north-east coast of Scotland beautiful but 'complacent', mirroring the conservatism and prejudice that played a part in her being given up for adoption. In contrast, when she visits her father's ancestral village, the experience is euphoric and is represented in terms of recognition and reunion. When she first approaches the village, she discovers 'a red dust road exactly like the one in my imagination . . . It's as if my footprints were already on the road before I even got there' (213). It is as if she and the land have already reciprocally marked each other:

> The earth is so copper warm and beautiful and the green of the long elephant grasses so lushly green they make me want to weep. I feel such a strong sense of affinity with the colours and the landscape, a strong sense of recognition. There's a feeling of liberation, and exhilaration, that at last, at last, at last I'm here. I feel shy with the landscape too, like I might be meeting a new blood relation. I almost feel like talking to it and whispering sweet nothings into its listening ear. (213)

The (lost) place is animated and figured as a flesh and blood relation, as a part of the self, an insight that resonates powerfully with an under-standing of epigenetic processes that mediate between the organism and the external environment. In this respect, Kay gestures towards the way in which epigenetic processes are being called on to explain both the positive and negative impacts of the external environment on the health of humans and other animals. As Lock suggests, this has become a particularly urgent issue in the context of changes to environments due to anthropogenic climate change.[41]

The fact that Kay finds herself—in both senses—on the red dust road is crucial, as the road becomes the governing metaphor for her narrative. Immediately prior to her father's refusal to see her, Kay thinks of Robert Frost's well-known poem 'The Road Not Taken', which is frequently invoked in adoption narratives to capture the sense the adoptee has of having switched from one life to another and being haunted, in consequence, by the life they could have led. Kay is aware that her life would have been very different if her father had taken on a paternal role, but she pulls back from these 'disquieting' speculations about her 'other self' and moves to detach herself from fantasies of ultimate origins and ends. Instead she defines her identity in terms of a tension between stability and change which is played out through 'the trail itself', in other words, through the interactions which shape the mutable subject:

> Perhaps it is the trail itself that is the interesting part, not the person at the end of it. At the end of the road, somewhere over the rainbow, a man is standing who turns out to be my father. He is small in height, sweating, dressed in a long white dress. He is not as I imagined. He is tiny. His big booming voice cannot hide the fact that he's full of hot air. (146)

For a range of reasons, Kay's father cannot meet the needs created by her experience of adoption, and the bathetic quality of her encounters with her birth parents reveals that the notion of an innate genetic connection is just so much 'hot air'. In place of this myth of origins she shapes a processual narrative in which the self is shown to unfold in and through complex, enmeshed inter-relations. Tracking the ways in which experience gets 'under the skin', *Red Dust Road* reveals the biological traces of individual and collective history, while in tracing

---

[41] See Margaret Lock, 'Recovering the Body', for a discussion of epigenetics in the era of the Anthropocene.

the subtle interactions between the individual and the environment, it emphasizes the porosity and unboundedness of the body. The memoir thus proposes an epigenetic ontogeny while at the same time suggesting an ontology that converges with Malabou's articulation of the wider philosophical implications of epigenetics.

As we have seen, epigenetics was initially embraced by scholars in the social sciences and the humanities because it promised freedom from the atomism and reductionism of neo-Darwinism. It opens the body up to society, identifying pathways whereby the psychosocial environment and the natural environment shape health and disease, and as environments are more amenable to intervention than genes, it is an area of research which is open to a progressive social agenda. However, as scholars of science and technology have argued, in its practical application, epigenetics is at risk of being assimilated to the instrumental logic which powered late-twentieth-century genetics and the HGP. Epigenetic research projects tend to locate the development of health and disease primarily at the level of individual bodies and behaviours, at the expense of structural views that encompass the social, political, and economic determinants of health. A great deal depends on how expansively or narrowly the environment is construed, an instructive case being the work on the epigenetic effects of maternal care in rat pups, mentioned above, in which the extrapolation from rats to humans is problematic in various ways. Most notably, the figure of the mother is taken in isolation and the many other factors which are important in development—for example peer relationships and the role of the father—are excluded from consideration.[42] A related issue is that as we have seen, if epigenetic marks are linked to specific health conditions, certain communities may be stigmatized. For example, research into the effects of psycho-social stress in low social and economic status households might be taken to suggest that the future is constrained by epigenetic marks acquired in early life, and that poverty and disadvantage are self-replicating. A similar concern arises in relation to research into the epigenetic effects of migration and

---

[42] For further discussion of these issues see Ruth Müller et al., 'The biosocial genome? Interdisciplinary perspectives on environmental epigenetics, health and society', *EMBO* 18 (October 2017) 1677–82, doi: 10.15252/embr.201744953.

trauma, which might lead to the stigmatization of refugees who are fleeing war and oppression.

Despite these concerns, the emancipatory potential of epigenetics has been emphasized by philosophers of biology, especially those who have resisted what they see as the limitations and distortions of neo-Darwinism. Evelyn Fox Keller, for example, sees discoveries in epigenetics as changing the very meaning of the genome, transforming it 'from an executive suite of directorial instructions to an exquisitely sensitive and reactive system that enables cells to regulate gene expression in response to their immediate environment'.[43] For John Dupré, epigenetics has been crucial in breaking the hold of the so-called Central Dogma and in demonstrating the diversity of forms of inheritance at the molecular level.[44] However, the most ambitious philosophical approach to epigenetics is that of Malabou, who aims to harness the insights of contemporary biology in order to transform philosophy itself.[45] As we have seen, for her the importance of epigenetics lies in the fact that it demonstrates the porous boundary between the organism and the environment and shows that the living being does not simply perform a programme, rather, 'the structure of the living being is an intersection between a given and a construction'.[46] Developing this line of enquiry, she re-reads the Kantian transcendental in

[43] Keller, 'The Postgenomic Genome', p. 10.

[44] Crick used this term to refer to the transfer of sequential information from DNA to RNA to protein in F.H.C. Crick, 'On Protein Synthesis', *Symp. Soc. Exp. Biol.*, XII (1958), 139–63 and the 'dogma' was re-stated in an article in *Nature* in 1970. In his autobiography, Crick acknowledges his misuse of the term, commenting that 'the use of the word dogma caused almost more trouble than it was worth. Many years later Jacques Monod pointed out to me that I did not appear to understand the correct use of the word dogma, which is a belief *that cannot be doubted*. I did apprehend this in a vague sort of way but since I thought that *all* religious beliefs were without foundation, I used the word the way I myself thought about it, not as most of the world does, and simply applied it to a grand hypothesis that, however plausible, had little direct experimental support.' See Francis Crick, *What Mad Pursuit: A Personal View of Scientific Discovery* (New York: Basic Books, 1988), p. 109.

[45] For a detailed exploration of the relationship between Malabou's philosophy and the life sciences see James, '(N)europlasticity, Epigenesis and the Void'. As he writes, in *Before Tomorrow* Malabou co-articulates 'a biological and philosophical reinterpretation of epigenesis and of epigenetic temporal becoming' (12).

[46] Catherine Malabou, 'One Life Only: Biological Resistance, Political Resistance', trans. Carolyn Shread, *Critical Inquiry* 42, Spring 2016, 429–38, https://criticalinquiry. uchicago.edu/one_life_only/ [accessed 12 April 2019].

terms of epigenesis and epigenetics, arguing in *Before Tomorrow* that the a priori structure of thought is folded into epigenetic development as 'meaning constructs itself, engenders itself—through epigenesis—it *becomes* what it is, starting from a blank'.[47] As this suggests, for Malabou, the transcendental rests on a void, as does the material substance of the living being, and it is their common exposure to the ontological void that enables the exchangeability of biological form and the transcendence of thought. While *The Peppered Moth* moves towards an intuition of the mutual imbrication of the biological and the symbolic, Kay's memoir is imbued with a profound understanding of the exchangeability of the two domains. As she explains in an interview about *Red Dust Road*, she thinks of her own origins in terms of a complexly entangled family tree, commenting that:

> as a society we are very interested in tracing family trees through biology and genes. But there are other family trees that preoccupy me. My adoptive mum always told my brother and I the story of our adoption, and listening to her tell me stories was one of the things that made me a writer, because she has such a vivid and dramatic way of talking. It seems that the story itself was handed down, and that for my mum and me, stories being passed down are as important as blood and genes. Red Dust Road is really about what makes us who we are, nature, nurture. I think of myself as being made from a mixture of porridge and myth! I think that if I were making a family tree, it would have my adoptive parents on it and their parents too, and it would be more complicated and intricate than the straightforward biological blood line.[48]

The metaphor of the tree, with its echoes of the biblical tree of life, the Darwinian tree, and the family tree, is here repurposed to signal the exchangeability of the biological and the symbolic. With the allusion to porridge and myth, Kay points to the equivalence of the materiality of being (porridge) and the transcendence of thought (myth), while her recursive, looping sentences, enact the mutually transformative relations of the biological and the social as they entwine to form a single tree.

---

[47] Catherine Malabou, *Before Tomorrow: Epigenesis and Rationality*, trans. Carolyn Shread (Cambridge: Polity Press, 2016), p. 98.

[48] Jackie Kay, book group interview, at http://www.bookgroup.info/041205/interview.php?id=73 [accessed 18 January 2014].

BIBLIOGRAPHY

Unpublished Sources

Harry Ransom Center, University of Texas at Austin
Kazuo Ishiguro Archive
Box 3.1 Ishiguro Never Let Me Go—'Rough Papers' (1 of 15) Feb. 2001—July 2003
Box 5.4 Ishiguro Never Let Me Go: 'Ideas as they Come', Notebook 1, Jan. '01–Sept. '02, 22/2/01
Box 5.4 Ishiguro Never Let Me Go: 'Ideas as they Come', Notebook 1, Jan. '01–Sept. '02, 14/1/02
Box 5.4 Ishiguro Never Let Me Go: 'Ideas as they Come', Notebook 1, Jan. '01–Sept. '02, 21/3/01
Box 5.4 Ishiguro Never Let Me Go: 'Ideas as they Come', Notebook 1, Jan. '01–Sept '02, loose sheet dated 2001
Box 5.6 Ishiguro Never Let me Go; 'First Rough Draft' Notebook—Clones 1' 2000–2001, 6/3/00, p. 5
Box 17.3 Ishiguro The Remains of the Day—'Butler notes and "ideas as they come"' 1986 –1987
Ian McEwan Archive
Green Notebook for *Enduring Love*, September 1995–January 1997, Container 7.10, unnumbered page
Saturday (novel 2005) Research. Surgery observation notes with Dr Neil Kitchen, medical articles, Ray Dolan email printouts, notes, 2004, Container 14.3
McEwan, Ian, Talk on Consciousness, Wellcome Trust, 2 October 2008, Container 28.4
Notes regarding Tony Blair (and plot notes for novel Saturday?), 8 May 2005, Box 29.3
Typescript for Radio 4 talk (untitled), c 1998, Box 29.3, 2

Published Works

Agamben, Giorgio, *The Coming Community*, trans. by Michael Hardt (Minneapolis: University of Minnesota Press, 1993)
Agamben, Giorgio, *Homo Sacer: Sovereign Power and Bare Life*, trans. by Daniel Heller-Roazen (Stanford: Stanford University Press, 1998)

Agamben, Giorgio, *The Open: Man and Animal*, trans. by Kevin Attell (Stanford: Stanford University Press, 2004)

Agamben, Giorgio, 'We Refugees', trans. by Michael Rocke, *Symposium*, 49:2 (1995), Periodicals Archive Online 116 rhttps://thehubedu-production.s3.amazonaws.com/uploads/1836/1e78843In tho-c11e-4036-8251-5406847cd 504/AgambenWeRefugees.pdf [accessed 14 Apr. 2019]

Alden, Natasha, 'Words of War, War of Words: *Atonement* and the Question of Plagiarism', in *Ian McEwan*, 2nd edn, ed. by Sebastian Groes (London: Bloomsbury, 2013), pp. 57–69

Allen, Elizabeth et al., 'Against "Sociobiology"', *The New York Review of Books*, 13 (1975) http://www.nybooks.com/articles/1975/11/13/against-sociobiology/ [accessed 28 May 2017]

Ardrey, Robert, *The Territorial Imperative: A Personal Inquiry into the Animal Origins of Property and Nations* (New York: Athenaeum, 1966)

Arendt, Hannah, 'The Decline of the Nation-State and the End of the Rights of Man', in *The Origins of Totalitarianism* (London: Penguin, 2017), pp. 349–96

Armstrong, Nancy, 'The Affective Turn in Contemporary Fiction', *Contemporary Literature* 55:3 (2014), 441–65

Arnold, Matthew, *Culture and Anarchy*, ed. J. Dover Wilson (Cambridge: Cambridge University Press, 1971)

Atwood, Margaret, '*The Handmaid's Tale* and *Oryx and Crake* in context', *PMLA* 119 (2004), 513–7. doi: 10.1632/003081204X20578

Banville, John, 'A Day in the Life', *New York Review of Books* (26 May 2005) http://www.nybooks.com/articles/2005/05/26/a-day-in-the-life/ [accessed 25 March 2018]

Barash, David, *The Whisperings Within: Evolution and the Origins of Human Nature* (London: Penguin, 1979)

Barkow, Jerome H., Lena Cosmides, and John Tooby, eds, *The Adapted Mind: Evolutionary Psychology and the Generation of Culture* (New York: Oxford University Press, 1992)

Baudrillard, Jean, *The Vital Illusion* (The Wellek Library Lectures), ed. by Julia Witwer (New York: Columbia University Press, 2001)

Beer, Gillian, *Darwin's Plots: Evolutionary Narrative in Darwin, George Eliot and Nineteenth-Century Fiction*, 2nd edn (Cambridge: Cambridge University Press, 2000)

Beer, Gillian, *Open Fields: Science in Cultural Encounter* (Oxford: Oxford University Press, 1996)

Bergson, Henri, *Creative Evolution* (trans. by Arthur Mitchell (London: Macmillan, 1911, 1960 printing)

Bhandar, Brenna and Jonathan Goldberg-Hiller, eds, *Plastic Materialities: Politics, Legality and Metamorphosis in the Work of Catherine Malabou* (Durham and London: Duke University Press, 2015)

Bingham, Marcus A., and Suzanne Simard. 'Ectomycorrhizal networks of Pseudotsuga menziesii var. glauca trees facilitate establishment of conspecific seedlings under drought.' *Ecosystems* 15:2 (2012): 188–99. doi: 10.1007/s10021-011-9502-2 at http://link.springer.com/article/10.1007%2Fs10021-011-9502-2 [accessed 15 Jan. 2014]

Bowler, Peter J., *Science for All: The Popularisation of Science in Early Twentieth-Century Britain* (Chicago: Chicago University Press, 2009)

Boyd, Brian, 'Evolutionary Theories of Art', in *The Literary Animal: Evolution and the Nature of Narrative*, ed. by Jonathan Gottschall and David Sloan Wilson (Evanston: Northwestern University Press, 2005), pp. 147–76

Bradotti, Rosi, *The Posthuman* (Cambridge: Polity Press, 2013)

Brockman, John, 'The Third Culture', https://www.edge.org/conversation/the-emerging [accessed 24 September 2017]

Brockman, John, *The Third Culture: Beyond the Scientific Revolution* (New York: Simon and Schuster, 1991)

Brown, Alistair, 'Uniting the Two Cultures of Body and Mind, in A.S. Byatt 'The Whistling Woman', *Literature and Science* 1:1 (2007), 55–72

Brown, Andrew, *The Darwin Wars; the Scientific Battle for the Soul of Man* (New York: Pocket Books, 2000)

Byatt, A.S., *Babel Tower* (London; Vintage, 2003)

Byatt, A.S., *The Biographer's Tale* (London; Vintage, 2001)

Byatt, A.S., 'Faith in Fiction', *Prospect Magazine* 20 November 2000, https://www.prospectmagazine.co.uk/magazine/faithinscience [accessed 26 Nov. 2018]

Byatt, A.S., The Feeling Brain', *Prospect Magazine*, 20 June 2003, https://www.prospectmagazine.co.uk/magazine/thefeelingbrain [accessed 30 Nov. 2018]

Byatt, A.S., 'Feeling Thought: Donne and the Embodied Mind', in *The Cambridge Companion to John Donne*, ed. by Achsah Guibbory (Cambridge: Cambridge University Press, 2006) pp. 247–58

Byatt, A.S., 'Fiction informed by science', *Nature* 435 (2005), 294–7

Byatt, A.S., *On Histories and Stories: Selected Essays* (London: Chatto & Windus, 2000)

Byatt, A.S., 'People in Paper Houses: Attitudes to "Realism" and "Experiment" in English Post-war Fiction', in *Passions of the Mind: Selected Writings* (1991) (London: Vintage, 1993), p. 167

Byatt, A.S., *Possession* (London: Vintage, 1991)

Byatt, A.S., 'Soul Searching', *Guardian* 14 February 2004, https://www.theguardian.com/books/2004/feb/14/fiction.philosophy [accessed 28 Nov. 2018]

Byatt, A.S., *Still Life* (London: Vintage, 1995)

Byatt, A.S., *The Virgin in the Garden* (London: Penguin, 1981)

Byatt, A.S., *A Whistling Woman* (2002) (London: Vintage, 2003)

Cameron, Deborah, 'Back to Nature', in *The Trouble & Strife Reader*, ed. by Deborah Cameron and Joan Scanlon (London: Bloomsbury, 2009), pp. 149–58

Carey, Nessa, *The Epigenetics Revolution: How Modern Biology is Rewriting Our Understanding of Genetics, Disease and Inheritance* (London: Icon Books, 2012)

Carson, Rachel, *Silent Spring* (London: Penguin, 2000)

Chakrabarty, Dipesh, 'The Climate of History: Four Theses', *Critical Inquiry* 35 (2009), 197–222

Chalmers, David, *The Character of Consciousness* (Philosophy of Mind) (Oxford: Oxford University Press, 2010)

Champagne, Frances A., 'Epigenetic influence of social experiences across the lifespan', *Developmental Psychobiology* 52:4 (2010), 299–311, doi: 10.1002/dev.20436

Changeux, Jean-Pierre, *Neuronal Man: The Biology of Mind*, new edn, trans. by Laurence Galey (Princeton: Princeton University Press, 1997)

Cherfas, Jeremy and John Gribbin, *The Redundant Male: Is Sex Irrelevant in the Modern World?* (London, Sydney, Toronto: Bodley Head, 1984)

Chodorow, Nancy, *The Reproduction of Mothering: Psychoanalysis and the Sociology of Gender*, second revised edn (Berkeley: University of California Press, 1999)

Cobb, Matthew, 'Happy 100th Birthday, Francis Crick (1916–2004)', Why Evolution is True website, available at https://whyevolutionistrue. wordpress.com/2016/06/08/happy-100th-birthday-francis-crick-1916-2004/ [accessed 17 July 2019]

Colebrook, Claire and Jason Maxwell, *Agamben*, Key Contemporary Thinkers (Cambridge: Polity Press, 2016)

Coole, Diana and Samantha Frost, eds, *New Materialisms: Ontology, Agency and Politics* (Durham and London: Duke University Press, 2010)

Cosmides, Lena and John Tooby, 'Does Beauty Build Adapted Minds? Toward an Evolutionary Theory of Aesthetics, Fiction and the Arts', *SubStance* 94:95 (2001), 24–5

Cosmides, Lena and John Tooby, 'The Psychological Foundations of Culture', in *The Adapted Mind: Evolutionary Psychology and the Generation of Culture*, ed. by Jerome H. Barkow, Leda Cosmides, and John Tooby (New York: Oxford University Press, 1992), pp. 19–136

Crick, Francis H.C., 'Discussion: Eugenics and Genetics', in *Man and His Future: A Ciba Foundation Volume*, ed. by Gordon Wolstenholme (Boston: Little, Brown & Co., 1963)

Crick, Francis H.C., 'On Protein Synthesis', *Symp. Soc. Exp. Biol.*, XII (1958), 139–63

Crick, Francis H.C., *What Mad Pursuit: A Personal View of Scientific Discovery* (New York: Basic Books, 1988)

Crossland, Rachel, *Modernist Physics: Waves, Particle and Relativities in the Writings of Virginia Woolf and D.H. Lawrence* (Oxford English Monographs) (Oxford: Oxford University Press, 2018)

Cyrulnik, Boris, *Resilience: How Your Inner Strength Can Set You Free from The Past*, trans. by David Macey (London: Penguin, 2009)

Damasio, Antonio, *Descartes' Error: Emotion, Reason and the Human Brain* (London: Vintage, 2006)

Darwin, Charles, *On the Origin of Species*, ed. with an introduction and notes by Gillian Beer (Oxford: Oxford World's Classics, 2008)

Darwin, Charles, letter to Asa Gray, 2 April 1860, Darwin Correspondence Project, https://www.darwinproject.ac.uk/letter/DCP-LETT-2743.xml [accessed 28 Nov 2018]

Daston, Lorraine and Peter Galison, *Objectivity* (Brooklyn, New York: Zone Books, 2007)

Dawkins, Richard, *The Extended Phenotype*, with an afterword by Daniel Dennett (Oxford: Oxford University Press, 1999)

Dawkins, Richard, 'Science, Delusion and the Appetite for Wonder', the Richard Dimbleby Lecture, https://www.edge.org/conversation/richard_dawkins-science-delusion-and-the-appetite-for-wonder, com [accessed 22 March 2018]

Dawkins, Richard, *The Selfish Gene*, 2nd edn (Oxford: Oxford University Press, 1989)

Derrida, Jacques, *Of Grammatology*, trans. Gayatri Chakravorty Spivak (Baltimore and London: Johns Hopkins University Press, 1976)

de Waal, Frans, 'Anthropomorphism and Anthropodenial: Consistency in Our Thinking about Humans and Other Animals', *Philosophical Topics* 27 (1999), 255–80

de Waal, Frans, *Our Inner Ape: The Best and Worst of Human Nature* (London: Granta Books, 2006)

Deichmann, Ute, *Biologists under Hitler*, trans. by Thomas Dunlop (Cambridge, MA: Harvard University Press, 1996)

Dennett, Daniel C., *Darwin's Dangerous Idea: Evolution and the Meanings of Life* ((London: Penguin, 1996)

Dissanayake, Ellen, *Art and Intimacy: How the Arts Began* (Seattle: University of Washington Press, 2000)

Dissanayake, Ellen, *What is Art For* (Seattle: University of Washington Press, 1988)

Doolittle, W. Ford, 'Is Junk DNA Bunk? A Critique of ENCODE,' *Proceedings of the National Academy of Sciences* 110 (2013), 5294–300. doi:10.1073/pnas.1221376110

Drabble, Margaret, *The Pattern in the Carpet: A Personal History with Jigsaws* (London: Atlantic Books, 2009)

Drabble, Margaret, *The Peppered Moth* (London: Penguin, 2001)

Drury, Shadia B., *Leo Strauss and the American Right* (Basingstoke: Palgrave Macmillan, 1999)

Dupré, John, *Processes of Life: Essays in the Philosophy of Biology* (Oxford: Oxford University Press, 2012)

Dupuy, Jean-Pierre, *The Mechanization of the Mind: On the Origins of Cognitive Science*, trans. by M.B. DeBevoise (Princeton and Oxford: Princeton University Press, 2000)

ENCODE Project Consortium, 'An integrated encyclopedia of DNA elements in the human genome', *Nature* 489, 7414 (2012), 57–74.

Ettinger, Bracha, *The Matrixial Gaze* (Minneapolis: University of Minnesota Press, 2006)

Fausto-Sterling, Anne, 'The Bare Bones of Sex: Part 1—Sex and Gender', *Signs* 30 (2005), 1491–527

Fausto-Sterling, Anne, 'Beyond Difference: Feminism and Evolutionary Psychology', in *Alas Poor Darwin*, ed. by Hilary S. Rose and Steven Rose (London: Vintage, 2001), pp. 174–89

Fausto-Sterling, Anne, *Myths of Gender: Biological Theories about Men and Women* (New York: Basic Books, second revised edn, 1992)

Fausto-Sterling, Anne, *Sex/Gender: Biology in a Social World* (Abingdon: Routledge, 2012)

Ferguson, Sam, 'Why Does Life-Writing Talk About Science? Foucault, Rousseau and the Early *Journal Intime*, *Biography* 40 (2017), 307–22

Foucault, Michel, *The Hermeneutics of the Subject: Lectures at the Collège de France*, trans. by Graham Burchell (New York: Picador, 2005)

Foucault, Michel, *The Order of Things: An Archaeology of the Human Sciences* (Abingdon: Routledge Classics, 2002)

Foucault, Michel, *Society Must Be Defended: Lectures at the Collège de France, 1975–76*, trans. by David Macey (London: Penguin, 2004)

Franklin, Sarah, *Dolly Mixtures: The Remaking of Genealogy* (Durham and London: Duke University Press, 2007)

Franklin, Sarah, 'Rethinking Reproductive Politics in Time, and Time in UK Reproductive Politics: 1978–2008', *Journal of the Royal Anthropological Institute (NS)* 20 (2014), 109–25

Freud, Sigmund, *The Standard Edition of the Complete Psychological Works of Sigmund Freud*, vol. 18, ed. by James Strachey (London: Hogarth Press and the Institute of Psychoanalysis, 1953–74)

Freud, Sigmund, 'The Theme of the Three Caskets', in The Penguin Freud Vol. 14, *Art and Literature* (Harmondsworth: Penguin, 1990), 234–47

Galton, Francis, *Inquiries into Human Faculty and its Development* (London: J.M. Dent & Sons, 1943)

Garrard, Greg, 'Ian McEwan's Next Novel and the Future of Ecocriticism', *Contemporary Literature* 50 (2009), 695–720

Garton Ash, Timothy, *Free World: Why a Crisis of the West Reveals the Opportunity of Our Time* (London: Penguin, 2005)

Gavron, Hannah, *The Captive Wife: Conflicts of Housebound Mothers* (London: Routledge and Kegan Paul, 1966)

Gilbert, Walter, 'Vision of the Grail', in *The Code of Codes*, ed. by D.J. Kevles and L. Hood (Cambridge: Harvard University Press, 1992), pp. 83–97

Gill, Josie, 'Science and Fiction in Zadie Smith's White Teeth', *Journal of Literature and Science* 6 (2013), 17–28

Gill, Josie, 'Written on the Face: Race and Expression in Kazuo Ishiguro's *Never Let Me Go*', *Modern Fiction Studies* 60:4 (2014), 844–62

Gilman, Charlotte Perkins, *Herland* (London: The Women's Press, 1979)

Gould, Stephen, Jay, 'More Things in Heaven and Earth', in *Alas Poor Darwin: Arguments Against Evolutionary Psychology*, ed. by Hilary Rose and Steven Rose (London: Vintage, 2001), 85–105

Gould, Stephen, Jay, *Rocks of Ages: Science and Religion in the Fullness of Life* (London: Vintage, 2002)

Gould, Stephen J., 'Sociobiology: the art of storytelling', *New Scientist* (16 November 1978), 530–3

Grosz, Elizabeth, *Becoming Undone: Darwinian Reflections on Life, Politics and Art* (Durham, NC: Duke University Press, 2011)

Guillory, John, 'The Sokal Affair and the History of Criticism', *Critical Inquiry* 28 (2002), 470–508

Hacking, Ian, 'Making Up People', *London Review of Books* 28 (17 Aug. 2006), 16, https://www.lrb.co.uk/v28/n16/ian-hacking/making-up-people [accessed 4 Nov. 2018]

Hacking, Ian, *The Taming of Chance* (Cambridge: Cambridge University Press, 1990)

Hamilton, William D., *Narrow Roads of Gene Land: The Collected Papers of W.D. Hamilton, Vol. 1, Evolution of Social Behaviour* (Oxford: Oxford University Press, 1998)

Hanson, Clare, *Eugenics, Literature and Culture in Post-war Britain* (London: Routledge, 2012)

Harvey, David, 'Neoliberalism as Creative Destruction', *Annals of the American Academy of Political and Social Science*, 610 (2007), 21–44

Hayles, N. Katherine, *How We Became Posthuman: Virtual Bodes in Cybernetics, Literature, and Informatics* (Chicago and London: Chicago University Press, 1999)

Head, Dominic, *Ian McEwan* (Manchester: Manchester University Press, 2013)

Herrnstein, Richard J. and Charles Murray, *The Bell Curve: Intelligence and Class Structure in American Life* (New York: Free Press, 1996)

Hirsch, Marianne, *The Generation of Postmemory: Writing and Visual Culture After the Holocaust* (New York: Colombia University Press, 2012)

Hoffman, Eva, *After Such Knowledge: A Meditation on the Aftermath of the Holocaust* (London: Vintage, 2005)

Hoffman, Eva, *The Secret* (London: Vintage, 2003)

Homans, Margaret, 'Adoption Narratives, Trauma, and Origins', *Narrative* 14 (2006), 4–26

Howell, Signe, *The Kinning of Foreigners: Transnational Adoption in a Global Perspective* (Oxford: Berghahn Books, 2007)

Hubbard, Ruth, *The Politics of Women's Biology* (New Brunswick, NJ: Rutgers University Press, 1990)

Huxley, T.H., Evolution and Ethics, Prolegomena, 1894, Project Gutenberg, https://www.gutenberg.org/files/2940/2940-h/2940-h.htm [accessed 17 July 2019]

Irigaray, Luce, *Speculum of the Other Woman*, trans. by Gillian C. Gill (Ithaca: Cornell University Press, 1985)

Ishiguro, Kazuo, *The Buried Giant* (London; Faber and Faber, 2016)

Ishiguro, Kazuo, *Never Let Me Go* (London: Faber and Faber Limited, 2006)

Ishiguro, Kazuo, Nobel acceptance speech, https://www.nobelprize.org/prizes/literature/2017/ishiguro/lecture/ [accessed 12 April 2019]

Ishiguro, Kazuo, *A Pale View of Hills* (London: Faber and Faber, 2005)

Ishiguro, Kazuo, *The Remains of the Day* (London: Faber and Faber, 1996)

Ishiguro, Kazuo, *The Unconsoled* (London: Faber and Faber, 2005)

Jablonka, Eva and Marion Lamb, *Evolution in Four Dimensions: Genetic, Epigenetic, Behavioural, and Symbolic Variation in the History of Life* (Cambridge, MA and London: MIT Press, 2005)

Jacob, Francois, *The Logic of Life: A History of Heredity*, trans. by Betty E. Spillmann (New York: Vintage Books, 1973)

James, Ian, '(N)europlasticity, Epigenesis and the Void', *parrhesia* 25 (2016), 1–19

Jasienska, Grazyna, 'Low birthweight of contemporary African Americans: An intergenerational effect of slavery?', *American Journal of Human Biology* 16 (2009), 16–24. doi:10.1002/ajhb.20824

Jones, Steve, *The Language of the Genes* (London: Flamingo, 1994)

Jones, Steve, *Y: The Descent of Men* (London: Little, Brown, 2002)

Josselyn, Sheena A., Stefan Kohler, and Paul W. Frankland, 'Heroes of the Engram', *Journal of Neuroscience* 37(2017), 4647–57, doi:10.1523/jneurosci.0056–17.2017

Kay, Jackie, *The Adoption Papers* (Newcastle upon Tyne: Bloodaxe Books, 1991)

Kay, Jackie, book group interview, at http://www.bookgroup.info/041205/interview.php?id=73 [accessed 18 Jan. 2014]

Kay, Jackie, *Red Dust Road* (London: Picador, 2010)

Kay, Lily E., *Who Wrote the Book of Life? A History of the Genetic Code* (Stanford: Stanford University Press, 2000)

Keen, Suzanne, 'Altruism Makes a Space for Empathy', branch collective, http://www.branchcollective.org/?ps_articles=suzanne-keen-altruism-makes-a-space-for-empathy-1852 [accessed 21 March 2018]

Kella, Elizabeth, 'Matrophobia and Uncanny Kinship: Eva Hoffman's *The Secret*', *Humanities*, 7:122 (2018), doi:10.3390/h7040122, available at https://www.mdpi.com/2076-0787/7/4/122/htm [accessed 13 April 2019]

Keller, Evelyn Fox, *The Century of the Gene* (Cambridge, MA: Harvard University Press, 2000)

Keller, Evelyn Fox, *A Feeling for the Organism: the Life and Work of Barbara McClintock* (New York: Holt, 2003)

Keller, Evelyn Fox, *Making Sense of Life: Explaining Biological Development with Models, Metaphors and Machines* (Cambridge, MA: Harvard University Press, 2002)

Keller, Evelyn Fox, *The Mirage of a Space between Nature and Nurture* (Durham and London: Duke University Press, 2010)

Keller, Evelyn Fox, 'The Postgenomic Genome', in Sarah S. Richardson and Hallam Stevens, eds, *Postgenomics: Perspectives on Biology and the Genome* (Durham and London: Duke University Press, 2015), pp. 9–31

Kennedy, Dane, *Islands of White: Settler Society and Culture in Kenya and Southern Rhodesia, 1890–1939* (Durham, NC: Duke University Press, 1987)

Kitcher, Philip, *Vaulting Ambition: Sociobiology and the Quest for Human Nature* (Cambridge, MA: MIT Press, 1987)

Kittay, Eva Feder, 'The Moral Harm of Migrant Carework: Realizing a Global Right to Care', *Philosophical Topics* 37 (2009), 53–73

Kolata, Gina, *Clone: The Road to Dolly and the Path Ahead* (London: Penguin, 1997)

Kristeva, Julia, *Black Sun: Depression and Melancholia* (New York: Columbia University Press, 1992)

Kuzawa, Chris and Elizabeth Sweet, 'Epigenetics and the Embodiment of Race: Developmental Origins of US Racial Disparities in Cardiovascular Health',

American Journal of Human Biology 21 (2009), 1–15. doi: 10.1002/ajhb.20822

Lacan, Jacques, *Four Fundamental Concepts in Psychoanalysis*, trans. by Alan Sheridan (London: Routledge, 2018)

Le Guin, Ursula K., 'Doris Lessing's First Sci-Fi Novel Reads Like a Debut Novel', *New Republic* (1979), https://newrepublic.com/article/115631/doris-lessing-shikasta-reviewed-ursula-le-guin [accessed 25 May 2017]

Le Guin, Ursula K., *The Wave in the Mind: Talks and Essays on the Writer, the Reader and the Imagination* (Boston, MA: Shambhala Publications Inc., 2004)

Leane, Elizabeth, 'Eggs, Emperors and Empire: Apsley Cherry-Garrard's "Worst Journey" as Imperial Quest Romance', *Kunapipi* 31 (2009)

Lessing, Doris, *Documents Relating to The Sentimental Agents in the Volyen Empire* (London: Flamingo, 1994)

Lessing, Doris, *The Golden Notebook* (London: Panther, 1972)

Lessing, Doris, *The Making of the Representative for Planet 8* (London: Jonathan Cape, 1982)

Lessing, Doris, *The Marriages Between Zones Three, Four and Five* (St Albans: Granada Publishing, 1981)

Lessing, Doris, 'Preface to the French Edition of *Seekers after Truth*', in *The Doris Lessing Reader* (New York: Alfred A. Knopf, 1989)

Lessing, Doris, *Putting the Questions Differently: Interviews with Doris Lessing, 1964–1994*, ed. by Earl G. Ingersoll (London: Flamingo, 1996)

Lessing, Doris, *Re: Colonised Planet 5 Shikasta* (St Albans: Granada Publishing, 1981)

Lessing, Doris, *The Sirian Experiments: the Report by Ambien 11, of the Five* (London: Flamingo, 1994)

Lessing, Doris, *A Small Personal Voice: Essays, Reviews, Interviews*, ed. by Paul Schlueter (New York: Alfred A Knopf, 1974)

Lessing, Doris, *Time Bites* (London: Harper Perennial, 2005)

Lessing, Doris, *Walking in the Shade: Volume Two of my Autobiography, 1949–1962* (London: Flamingo, 1998)

Levin, Ira, *The Boys from Brazil* (London: Corsair, 2011)

Lewontin, Richard, 'The Apportionment of Human Diversity', *Evolutionary Biology* 6 (1972), 381–98

Lock, Margaret, 'Comprehending the Body in the Era of the Epigenome', *Current Anthropology* 56 (2015), 151–77

Lock, Margaret, *Encounters with Aging: Mythologies of Aging in Japan and North America* (Berkeley: University of California Press, 1993)

Lock, Margaret, 'Mutable Environments and Permeable Human Bodies', *Journal of the Royal Anthropological Institute (NS)*, 24 (September 2018,) 449–74

Lock, Margaret, 'Recovering the Body', *Annual Review of Anthropology* 46 1–14, October 2017. doi: 10.1146/annurev-anthro-102116-041253, https://www.annualreviews.org/doi/full/10.1146/annurev-anthro-102116-041253 [accessed 13 May 2019]

Makinnon, Catharine, 'Sexuality, Pornography and Method: "Pleasure under Patriarchy"', *Ethics*, 99 (1989) 314–46

Malabou, Catherine, *Before Tomorrow: Epigenesis and Rationality*, trans. by Carolyn Shread (Cambridge: Polity Press, 2016)

Malabou, Catherine, 'Following Generation', trans. by Simon Porzak, *Qui Parle* 20:2 (2012), 19–33

Malabou, Catherine, *The Heidegger Change: On the Fantastic in Philosophy*, trans. by Peter Skafish (Albany, NY: SUNY Press, 2011)

Malabou, Catherine, *The New Wounded: From Neurosis to Brain Damage*, trans. by Steven Miller (New York; Fordham University Press, 2012)

Malabou, Catherine, 'One Life Only: Biological Resistance, Political Resistance', trans. by Carolyn Shread, *Critical Inquiry* 42 (2016), 429–38, available at https://criticalinquiry.uchicago.edu/one_life_only/ [accessed 12 Apr. 2019]

Malabou, Catherine, *Ontology of the Accident: An Essay on Destructive Plasticity*, trans. by Carolyn Shread (Cambridge: Polity Press, 2012)

Malabou, Catherine, *Plasticity at the Dusk of Writing: Dialectic, Destruction, Deconstruction*, trans. by Carolyn Shread (New York and Chichester: Columbia University Press, 2010)

Malabou, Catherine, *What Shall We Do With Our Brain?*, trans. by Sebastian Rand (New York: Fordham University Press, 2008)

Malik, Kenan, *Man, Beast and Zombie: The New Science of Human Nature* (New Brunswick: Rutgers University Press, 2002)

Marsh, Henry, 'Into the Grey Zone: can we really be conscious while in a coma?', *New Statesman* 27 August 2017, https://www.newstatesman.com/culture/books/2017/08/grey-zone-can-one-really-be-conscious-while-coma [accessed 14 Apr. 2019]

Martin, Paul, 'Toward a Biology of Social Experience?', response to Margaret Lock, 'Comprehending the Body in the Era of the Epigenome', *Current Anthropology* 56, 151–277 (pp. 167–8) (April 2015)

Matthews, Sean and Sebastian Groes, *Kazuo Ishiguro: Contemporary Critical Perspectives* (London: Continuum, 2010)

Maturana, Humberto R. and Francisco J. Varela, *Autopoiesis and Cognition: The Realization of the Living* (Dortrecht: D. Reidel, 1980)

Maturana, Humberto R. and Francisco J. Varela, *The Tree of Knowledge: The Biological Roots of Human Understanding* (1987), trans. by Robert Paolucci, revised edn. (Boston and London: Shambhala, 1992)

Maynard Smith, John, *The Evolution of Sex* (Cambridge: Cambridge University Press, 1978)

Mazumdar, Pauline M.H., *Eugenics, Human Genetics and Human Failings: The Eugenics Society, its Sources and its Critics in Britain* (London: Routledge, 1991)

McEwan, Ian, *Atonement* (London: Vintage, 2001)

McEwan, Ian, 'A Boot Room in the Frozen North', http://www.capefarewell.com/explore/215-a-boot-room-in-the-frozen-north.html [accessed 25 March 2018]

McEwan, Ian, *Enduring Love* (London: Vintage, 2006)

McEwan, Ian, 'Faith and doubt at ground zero', *Frontline*, available at https://www.pbs.org/wgbh/pages/frontline/shows/faith/interviews/mcewan.html [accessed 3 January 2018)]

McEwan, Ian, 'The Great Odyssey', *Guardian*, 'Saturday Review' (9 June 2001), 1–3

McEwan, Ian, 'Literature, Science and Human Nature', in *The Literary Animal: Evolution and the Nature of Narrative*, ed. by Jonathan Gottschall and David Sloan Wilson (Evanston: Northwestern University Press, 2005), pp. 5–19

McEwan, Ian, *Nutshell* (London: Jonathan Cape, 2016)

McEwan, Ian, 'On *Nutshell*', Guardian Books Podcast, 2 September 2016, https://www.theguardian.com/books/audio/2016/sep/02/ian-mcewan-on-his-novel-nutshell-books-podcast [accessed 25 March 2018]

McEwan, Ian, *Saturday* (London: Vintage, 2005)

McEwan, Ian, 'When faith in fiction falters—and how it is restored', *Guardian* 16 February 2013, https://www.theguardian.com/books/2013/feb/16/ian-mcewan-faith-fiction-falters

McGowan, Patrick O., Aya Sasaki, Ana C. D'alessio, Sergiy Dymov, Benoit Labonté, Moshe Szyf, Gustavo Turecki, and Michael J. Meaney. 'Epigenetic regulation of the glucocorticoid receptor in human brain associates with childhood abuse', *Nature Neuroscience* 12 (2009), 342–48. doi:10.1038/nn.2270

McGuinness, D. et al., 'Socio-economic status is associated with epigenetic differences in the pSoBid cohort', *International Journal of Epidemiology* 41 (9 Jan. 2012), 151–60 (doi:10.1093/ije/dyr215)

McKinnon, Susan, *Neo-liberal Genetics: the Myths and Moral Tales of Evolutionary Psychology* (Chicago: Prickly Paradigm Press, 2005)

McLeod, John, *Lifelines: Writing Transcultural Adoption* (London: Bloomsbury, 2015)

Meloni, Maurizio, response to Margaret Lock, 'Comprehending the Body in the Era of the Epigenome', *Current Anthropology* 56, 151–277 (p. 168) (April 2015)

Meloni, Maurizio, 'Epigenetics for the social sciences: justice, embodiment, and inheritance in the postgenomic age', *New Genetics and Society* 34 (2015), 125–51. doi:10.1080/14636778.2015.1034850

Meloni, Maurizio, 'Race in an Epigenetic Time: Thinking Biology in the Plural', *British Journal of Sociology* 68 (2017), 389–409, doi:10.1111/1468-4446.12248

Mills, Catherine, *The Philosophy of Agamben* (London: Routledge, 2008)

Milton, John, *Paradise Lost*, ed. by Alastair Fowler (London: Longman Annotated English Poets, 2nd edition, 2006)

Mitchison, Naomi, *Solution Three*, afterword by Susan Squier (New York: The Feminist Press at the City University of New York, 1975)

Morris, Desmond, *The Naked Ape: A Zoologist's Study of the Human Animal* (London: Jonathan Cape, 1967)

Morton, Stephen, *States of Emergency: Colonialism, Literature and Law* (Liverpool: Liverpool University Press, 2013)

Morton, Timothy, *Dark Ecology: For a Logic of Future Coexistence* (The Wellek Library Lectures), (New York: Columbia University Press, 2016)

Morton, Timothy, *Ecology without Nature* (Cambridge, MA: Harvard University Press, 2009)

Mullan, John, 'Positive feedback', *Guardian*, Saturday 1 April 2006, https://www.theguardian.com/books/2006/apr/01/kazuoishiguro [accessed 14 Apr. 2019]

Müller, Ruth, Clare Hanson, Mark Hanson, Michael Penkler, Georgia Samaras, Luca Chiapperino, John Dupré et al., 'The biosocial genome? Interdisciplinary perspectives on environmental epigenetics, health and society', *EMBO reports* 18 (2017), 1677–82. doi: 10.15252/embr.201744953

Müller-Wille, Stefan and Hans-Jörg Rheinberger, *A Cultural History of Heredity* (Chicago and London: Chicago University Press, 2012)

Myrdal, Alva and Viola Klein, *Women's Two Roles: Home and Work*, 2nd revised edn (London: Routledge and Kegan Paul, 1968)

Nagel, Thomas, *The View from Nowhere* (Oxford: Oxford University Press, 1989)

Nagel, Thomas, 'Why so Cross?', *London Review of Books*, 21:7 (1 April 1999), 22–3, https://www.lrb.co.uk/v21/n07/thomas-nagel/why-so-cross [accessed 25 September 2017]

Negri, Antonio, 'The Sacred Dilemma of Inoperosity: On Giorgio Agamben's *Opus Dei*, Unimode 2.0, 18/09/2012, http://www.uninomadaccessede.org/negri-on-agamben-opus-dei/ [accessed 21 July 2019]

Nelkin, Dorothy 'Less Selfish than Sacred? Genes and the Religious Impulse in Evolutionary Psychology', in *Alas Poor Darwin: Arguments Against*

Evolutionary Psychology, ed. by Hilary S. Rose and Steven Rose (London: Vintage, 2001), 14–27

Nelson, Charles A., Nathan A. Fox, and Charles H. Zeanah, *Romania's Abandoned Children: Deprivation, Brain Development, and the Struggle for Recovery* (Cambridge, MA: Harvard University Press, 2013)

Nesse, Randolph M., 'Why a Lot of People with Selfish Genes are Pretty Nice Except for their Hatred of *The Selfish Gene*', in *Richard Dawkins: How a Scientist Changed the Way We Think*, ed. by Alan Grafen and Mark Ridley (Oxford: Oxford University Press, 2006)

Niewöhner, Jorg, 'Epigenetics: Embedded Bodies and the Molecularisation of Biography and Milieu', *BioSocieties* 6:3 (2001), 279–98

Niewöhner, Jorg and Margaret Lock, 'Situating Local Biologies: Anthropological Perspectives on Environment/Human Entanglements', *BioSocieties* 13 (December 2018), 681–97 (p. 693)

Owen, Adrian, *Into the Grey Zone: A Neuroscientist Explains the Mysteries of the Brain and the Borders between Life and Death* (London: Guardian Faber Publishing, 2017)

Peluffo, Alexandre E., 'The "Genetic Program": Behind the Genesis of an Influential Metaphor', *Genetics* 200 (2015), 685–96. doi:10.1534/genetics.115.178418

Pembrey, Marcus E., Lars Olov Bygren, Gunnar Kaati, Sören Edvinsson, Kate Northstone, Michael Sjöström, and Jean Golding. 'Sex-specific, male-line transgenerational responses in humans', *European Journal of Human Genetics* 14 (2006), 159–66

Piattelli-Palmarini, Massimo, ed., *Language and Learning: The Debate between Jean Piaget and Noam Chomsky* (Cambridge, MA: Harvard University Press, 1980)

Pinker, Steven, *The Blank Slate: The Modern Denial of Human Nature* (London: Penguin, 2002)

Pinker, Steven, *How the Mind Works* (London: Penguin, 1997)

Pinker, Steven, *The Language Instinct: How the Mind Creates Language* (London: Penguin, 2015)

Radrouf, David, Antoine Luz, Diego Cosmelli, Jean-Philippe Lachaux, Michel le Van Quyen, 'From autopoiesis to neurophenomenology: Francisco Varela's exploration of the biophysics of being', *Biological Research* 36 (2003), 21–59

Richardson, Sarah and Hallam Stevens, *Postgenomics* (Durham and London: Duke University Press, 2015)

Ridley, Matt, *Genome: The Autobiography of a Species in 23 Chapters* (London: Fourth Estate, 2000)

Ridley, Matt, *The Origins of Virtue* (London: Penguin, 1997)

Roberts, Dorothy E., 'What's Wrong with Race-Based Medicine?': Genes, Drugs, and Health Disparities', *Minnesota Journal of Law, Science and Technology* 12 (2011), 1–21

Robinson, Marilynne, *Absence of Mind: the Dispelling of Inwardness from the Modern Myth of the Self* (New Haven and London: Yale University Press, 2010)

Robinson, Marilynne, *The Givenness of Things: Essays* (New York: Picador, 2016)

Roof, Judith, *The Poetics of DNA* (Minneapolis: University of Minnesota Press, 2007)

Rose, Hilary, 'Colonising the Social Sciences', in *Alas Poor Darwin: Arguments Against Evolutionary Psychology*, ed. by Hilary Rose and Steven Rose (London: Jonathon Cape, 2000), pp. 106–28

Rose, Hilary, 'Eugenics and Genetics: the Conjoint Twins?', *New Formations* 60 (Spring 2007), 13–26

Rose, Nikolas, *The Politics of Life Itself: Biomedicine, Power, and Subjectivity in the Twenty-First Century* (Princeton and Oxford: Princeton University Press, 2007)

Rose, Steven, R.C. Lewontin, and Leon J. Kamin, *Not in our Genes: Biology, Ideology and Human Nature* (London: Penguin, 1984)

Rowe, Marsha, 'If you mate a swan and a gender, who will ride?' in *Notebooks, Memoirs, Archives: Reading and Rereading Doris Lessing*, ed. by Jenny Taylor (London: Routledge and Kegan Paul 1982), pp. 191–205

Rushdie, Salman, *Imaginary Homelands: Essays and Criticism 1981–1991* (London: Penguin, 1991)

Sagan, Dorionn and Lynn Margulis, 'God, Gaia, and Biophilia', in *The Biophilia Hypothesis*, ed. by Stephen R. Kellert and Edward O. Wilson (Washington, DC: Island Press, 1993), pp. 345–64

Segerstråle, Ullica, 'An Eye on the Core: Dawkins and Sociobiology', in *Richard Dawkins: How a Scientist Changed the Way We Think*, ed. by Alan Grafen and Matt Ridley (Oxford: Oxford University Press, 2006), pp. 75–97

Shah, Idries, *The Sufis* (London: ISF Publishing, 2014)

Shaviro, Steven, *The Universe of Things: On Speculative Realism* (Minneapolis: University of Minnesota Press, 2014)

Shuttleworth, Sally, 'Life in the Zooniverse: Working with Citizen Science', *Journal of Literature and Science* 10 (2017), 46–51

Smith, Zadie, *White Teeth* (London: Penguin, 2001)

Spector, Tim, *Identically Different: How You Can Change Your Genes* (London: Weidenfeld and Nicholson, 2012)

Stacey, Jackie, *The Cinematic Life of the Gene* (Durham and London: Duke University Press, 2010)

Stengers, Isabelle, *Thinking with Whitehead: A Free and Wild Creation of Concepts*, trans. by Michael Chase (Cambridge, MA: Harvard University Press, 2011)

Strathausen, Carsten, *Bioaesthetics: Making Sense of Life in Science and the Arts* (Minneapolis: University of Minnesota Press, 2017)

Strathern, Marilyn, *After Nature: English Kinship in the Late Twentieth Century* (Cambridge: Cambridge University Press, 1992)

Sullivan, Shannon, *The Physiology of Sexist and Racist Oppression* (New York: Oxford University Press, 2015)

Todes, Daniel, 'Global Darwin: Contempt for Competition', *Nature* 462, 7269 (2009), 36. doi: 10.1038/462036a

Tooby, John and Leda Cosmides, 'The Evolution of War and its Cognitive Foundations', *Institute for Evolutionary Studies Technical Report 88–1* (1988), https://pdfs.semanticscholar.org/7f95/d9d117721df9e69b929b004d9d85ea6c5 60d.pdf [accessed 22 March 2018]

UNESCO and Its Programme, Vol. 3, The Race Question (Paris: UNESCO, 1950)

UNESCO, *Racism and Apartheid: South Africa and Southern Rhodesia* (Paris: The Unesco Press, 1975), 40. Available at http://unesdoc.unesco.org/images/ 0001/000161/016163eo.pdf [accessed 28 May 2017]

The U.S.–Venezuela Collaborative Research Project and Nancy S. Wexler, 'Venezuelan Kindreds Reveal That Genetic and Environmental Factors Modulate Huntington's Disease Age of Onset', *Proceedings of the National Academy of Sciences of the United States of America* 101 (2004): 3498–503. (PMC. Web. 29 May 2018, https://www.ncbi.nlm.nih.gov/pmc/articles/ PMC373491/ [accessed 29 May 2018]

van't Hof, Arjen E., Pascal Campagne, Daniel J. Rigden, Carl J. Yung, Jessica Lingley, Michael A. Quail, Neil Hall, Alistair C. Darby, and Ilik J. Saccheri, 'The industrial melanism mutation in British peppered moths is a transposable element', *Nature* 534, (2016) 102–5, doi: 10.1038/nature17951

Venter, C.J., 'Remarks at the Human Genome Announcement' *Functional and Integrative Genomics* (2000), 1: 154–5 (154), doi: 10.1007/s101420000026

Vitale, Francesco, 'The Text and the Living: Jacques Derrida between Biology and Deconstruction', *The Oxford Literary Review* 36:1 (2014), 95–114, doi: 10.3366/olr.2014.0089 ((101)

Waddington, Conrad H., *An Introduction to Modern Genetics* (London: George Allen and Unwin, 1939)

Waddington, Conrad H., *The Strategy of the Genes* (London: George Allen and Unwin, 1957)

Waddington, Conrad H., 'Whitehead and Modern Science', in John B. and David R. Griffin Cobb, Jr, *Mind in Nature: the Interface of Science and*

Philosophy, https://www.religion-online.org/book-chapter/chapter-5-white head-and-modern-science-by-c-h-waddington/ [accessed 30 August 2016]

Wardi, Dina, *Memorial Candles: Children of the Holocaust* (The International Library of Group Psychotherapy and Group Process) (London: Routledge, 1992)

Watson, James, *The Double Helix* (London: Penguin Books, 1999)

Watson, James D. and Francis H.C. Crick, 'Molecular Structure of Nucleic Acids: A Structure of Deoxyribose Nucleic Acid', *Nature* 171 (1953), 737–38

Waugh, Patricia, 'The Novel as Therapy: Ministrations of Voice in an Age of Risk', *Journal of the British Academy* 3 (2015), 35–68

Weaver, Ian C.G., Nadia Cervoni, Frances A. Champagne, Ana C. D'Alessio, Shakti Sharma, Jonathan R. Seckl, Sergiy Dymov, Moshe Szyf, and Michael J. Meaney, 'Epigenetic programming by maternal behavior', *Nature Neuroscience* 7:8 (2004): 847–54

Webster, Brenda, 'Conversation with Eva Hoffman', *Women's Studies: An Interdisciplinary Journal* 32 (2003), 761–9, doi: 10.1080/00497870390221927

Weinbaum, Alys Eve, 'Racial Aura: Walter Benjamin and the Work of Art in a Biotechnological Age', *Literature and Medicine* 26 (2007), 207–39

Weindling, Paul, *Health, Race and German Politics* (Cambridge: Cambridge University Press, 1993)

Weismann, August, *Essays upon Heredity* (Oxford: Oxford University Press, 1889), http://www.esp.org/books/weismann/essays/facsimile/contents/weismann-essays-1-a-fm.pdf, [accessed 27 Nov. 2018]

Wellman, H.M., *The Child's Theory of Mind* (Cambridge, MA: MIT Press, 1992)

Whitehead, Alfred North, *The Concept of Nature* (Cambridge: Cambridge University Press, 1920)

Whitehead, Alfred North, *Process and Reality* (New York: Free Press, 1978)

Whitehead, Anne, 'Writing with Care: Kazuo Ishiguro's *Never Let Me Go*', *Contemporary Literature* 52:1 (2011), 54–83

Wiener, Norbert, *The Human Use of Human Beings: Cybernetics and Society* (Boston, MA: Houghton Mifflen, 1950)

Williams, George C., 'Pleiotropy, Natural Selection and the Evolution of Senescence', *Evolution* 11:4 (1957), 398–411

Williams, George C., *Sex and Evolution* (Princeton and Oxford: Princeton University Press, 1975)

Wilmut, Ian, Keith Campbell and Colin Tudge, *The Second Creation: The Age of Biological Control by the Scientists Who Cloned Dolly* (London: Headline Book Publishing, 2000)

Wilson, Edward O., *Biophilia* (Cambridge, MA: Harvard University Press, 1984)

Wilson, Edward O., *Consilience: the Unity of Knowledge* (London: Abacus, 2013)

Wilson, Edward O., *On Human Nature*, with a new preface (Cambridge, MA: Harvard University Press, 2004)

Wilson, Edward O., *Naturalist* (Washington, DC: Island Press, 1994)

Wilson, Edward O., *Sociobiology: The New Synthesis* (Cambridge, MA: Harvar University Press, 1975)

Wilson, Elizabeth A, *Gut Feminism* (Durham, NC: Duke University Press, 2015)

Winterson, Jeanette, *Written on the Body* (London: Jonathan Cape, 1992)

Wohlleben, Peter, *The Hidden Life of Trees: What They Feel, How They Communicate: Discoveries from a Secret World*, trans. by Jane Billinghurst (London: William Collins, 2017)

Wood, James, 'Containment: Trauma and Manipulation in Ian McEwan', in *The Fun Stuff and Other* Essays (London: Vintage, 2014), p. 185

Woolf, Virginia, 'Modern Fiction', 1921, http://gutenberg.net.au/ebooks03/0300031h.html#C12 [accessed 25 March 2018]

Wright, Robert, *The Moral Animal: Why We Are the Way We Are* (London: Abacus, 1996)

INDEX

For the benefit of digital users, indexed terms that span two pages (e.g., 52–53) may, on occasion, appear on only one of those pages.